SpringerBriefs in History of Science and Technology

Series Editors

Gerard Alberts, University of Amsterdam, Amsterdam, The Netherlands

Theodore Arabatzis, University of Athens, Athens, Greece

Bretislav Friedrich, Fritz Haber Institut der Max Planck Gesellschaft, Berlin, Germany

Ulf Hashagen, Deutsches Museum, Munich, Germany

Dieter Hoffmann, Max-Planck-Institute for the History of Science, Berlin, Germany

Simon Mitton, University of Cambridge, Cambridge, UK

David Pantalony, Ingenium - Canada's Museums of Science and Innovation / University of Ottawa, Ottawa, Canada

Matteo Valleriani, Max-Planck-Institute for the History of Science, Berlin, Germany

The *SpringerBriefs in the History of Science and Technology* series addresses, in the broadest sense, the history of man's empirical and theoretical understanding of Nature and Technology, and the processes and people involved in acquiring this understanding. The series provides a forum for shorter works that escape the traditional book model. SpringerBriefs are typically between 50 and 125 pages in length (max. ca. 50.000 words); between the limit of a journal review article and a conventional book.

Authored by science and technology historians and scientists across physics, chemistry, biology, medicine, mathematics, astronomy, technology, and related disciplines, the volumes will comprise:

1. Accounts of the development of scientific ideas at any pertinent stage in history: from the earliest observations of Babylonian Astronomers, through the abstract and practical advances of Classical Antiquity, the scientific revolution of the Age of Reason, to the fast-moving progress seen in modern R&D;
2. Biographies, full or partial, of key thinkers and science and technology pioneers;
3. Historical documents such as letters, manuscripts, or reports, together with annotation and analysis;
4. Works addressing social aspects of science and technology history (the role of institutes and societies, the interaction of science and politics, historical and political epistemology);
5. Works in the emerging field of computational history.

The series is aimed at a wide audience of academic scientists and historians, but many of the volumes will also appeal to general readers interested in the evolution of scientific ideas, in the relation between science and technology, and in the role technology shaped our world.

All proposals will be considered.

Alica-Nana Citron

Spinning the Cosmos

Volvelles in the Early Modern Commentary Tradition of Johannes de Sacrobosco's *De Sphaera*

Alica-Nana Citron
Max Planck Institute for the History
of Science
Berlin, Germany

University of Oslo
Oslo, Norway

ISSN 2211-4564　　　　　　　ISSN 2211-4572　(electronic)
SpringerBriefs in History of Science and Technology
ISBN 978-3-031-90975-7　　　ISBN 978-3-031-90976-4　(eBook)
https://doi.org/10.1007/978-3-031-90976-4

© The Author(s) 2025. This book is an open access publication.

Open Access This book is licensed under the terms of the Creative Commons Attribution 4.0 International License (http://creativecommons.org/licenses/by/4.0/), which permits use, sharing, adaptation, distribution, and reproduction in any medium or format, as long as you give appropriate credit to the original author(s) and the source, provide a link to the Creative Commons license and indicate if changes were made.
The images or other third party material in this book are included in the book's Creative Commons license, unless indicated otherwise in a credit line to the material. If material is not included in the book's Creative Commons license and your intended use is not permitted by statutory regulation or exceeds the permitted use, you will need to obtain permission directly from the copyright holder.
The use of general descriptive names, registered names, trademarks, service marks, etc. in this publication does not imply, even in the absence of a specific statement, that such names are exempt from the relevant protective laws and regulations and therefore free for general use.
The publisher, the authors, and the editors are safe to assume that the advice and information in this book are believed to be true and accurate at the date of publication. Neither the publisher nor the authors or the editors give a warranty, expressed or implied, with respect to the material contained herein or for any errors or omissions that may have been made. The publisher remains neutral with regard to jurisdictional claims in published maps and institutional affiliations.

This Springer imprint is published by the registered company Springer Nature Switzerland AG
The registered company address is: Gewerbestrasse 11, 6330 Cham, Switzerland

If disposing of this product, please recycle the paper.

Acknowledgements

This work began as my final graduate thesis at the Technische Universität, Berlin (Germany). The present book is not just a monograph to me, but the outcome of years of learning, training, knowledge acquisition, meeting wonderful people, and immense personal progress. I was guided through the process of writing this book and especially the research beforehand with the support of several people. First of all, my thanks go to Prof. Dr. Matteo Valleriani. Matteo not only brought me to the Max Planck Institute for the History of Science in 2017 to work with him on his wonderful project about the *Sphaera* of Johannes de Sacrobosco, but taught me an enormous amount about academia, scientific practice, and the early modern period. He suggested the topic for this work and answered all my questions with knowledge and dedication, providing an immeasurable amount of support and guidance along the way as my supervisor. I will be forever grateful I met him. He is not just an impressive researcher but also one of the most warm-hearted people I've ever met.

Prof. Dr. Friedrich Steinle also deserves my gratitude. Throughout my studies at the Technische Universität I found him to be a supportive researcher, acknowledging the potential of his students and guiding them through the academic process with a smile and helpful feedback. Having him as my professor created a welcoming environment for my first steps in the Humanities. Both Matteo and Friedrich inspired me to make a career in academia.

My second supervisor, Prof. Dr. Günther Oestmann, not only agreed to support my topic and assist me with his wealth of knowledge in an advisory capacity, but also provided very helpful feedback on all kinds of questions. Besides that, he is a kind and caring individual beyond the academic sphere.

Furthermore, I would like to thank the *Sphaera* team, but especially Victoria Beyer and Noga Shlomi, for finding volvelle parts during the process of image clustering and discussing with me the differences between image and instrument. Additionally, their presence and friendship made working at the Max Planck Institute one of the best experiences I ever had. The database, the clustering, and my research were made possible by the project's IT support, predominantly provided by Hassan El-Hajj, Anika Merklein, Olya Nicolaeva, Daan Lockhorst, and Jochen Büttner. I am especially grateful to Hassan and Anika for their support and for the extraction of the volvelle data for my research, providing help with any further questions on data extractions, and the ongoing technical support.

I would like to extend my thanks, too, to Prof. Dr. Marius Buning. He provided extremely helpful feedback and support during my first steps in my Ph.D. program, supported the idea of publishing this work from its inception, and trusted me to finish this book project in parallel with my Ph.D., which amounts to a great leap of trust.

My research was supported by a scholarship from the German Academic Scholarship Foundation (Studienstiftung des deutschen Volkes), for which I am extremely grateful.

Finally, I want to thank my husband, Dennis, and our kids, as well as my Paps. My Paps helped me work on this book by keeping me company in his wonderful house and providing me with food, just by sitting next to me and doing crossword puzzles, which filled me with ideas and serenity.

My husband Dennis' immense support from the very first second we fell in love enabled me to achieve things that would have been impossible before. I can't express in words how meeting him changed my life for the better.

A. and T.: Thank you for making me laugh, filling my heart with joy, and teaching me things I would have never known before bringing you into this world.

Funded by the European Union (ERC, BE4COPY, 101042034). The views and opinions expressed in the content are solely those of the authors and do not necessarily reflect the views of the European Union or the European Research Council. The European Union and the granting authority cannot be held accountable for the expressed views and opinions.

Contents

1 **Paper Instruments as Objects of Research** 1
  Bibliography 4

2 **Interactive Prints and Paper Instruments** 7
  2.1 Astronomical Instruments 13
  2.2 Stereographic Projection 14
  Bibliography 17

3 **The *Sphaera* of Johannes de Sacrobosco in the Early Modern Period** 21
  3.1 Knowledge Within the *Sphaera* 23
  3.2 The Sphere—A Research Project 28
  Bibliography 31

4 **Volvelles in the *Sphaera* Corpus** 35
  4.1 The Function of Volvelles 36
  4.2 Situation Report: Volvelles 37
    4.2.1 Data Collection 39
    4.2.2 Volvelle Groups 40
    4.2.3 *Privilegia impressoria* in the Context of the *Sphaera* 44
  4.3 The Wittenberg Group 47
    4.3.1 Base Discs and Parts 48
    4.3.2 Correctly Assembled Volvelles 51
    4.3.3 Incorrectly Assembled Volvelles 53
  Bibliography 53

5 **The Material Culture of the Wittenberg Group** 57
  5.1 Method: The Winterthur Model 58
  5.2 Thing Knowledge 61
  Bibliography 62

6	**The Function and Use of Volvelles from the Wittenberg Group**		63
	6.1 Eclipse Volvelle		66
		6.1.1 Identification	66
		6.1.2 Evaluation	70
		6.1.3 Cultural Analysis	70
	6.2 Horizon Volvelle		71
		6.2.1 Identification	71
		6.2.2 Evaluation	75
		6.2.3 Cultural Analysis	75
		6.2.4 Interpretation	75
	6.3 Zodiac Volvelle		75
		6.3.1 Identification	76
		6.3.2 Evaluation	78
		6.3.3 Cultural Analysis	79
		6.3.4 Interpretation	80
	6.4 Heliacal Volvelle		80
		6.4.1 Identification	80
		6.4.2 Evaluation	86
		6.4.3 Cultural Analysis	86
		6.4.4 Interpretation	86
	Bibliography		87
7	**Conclusion**		91
	Bibliography		97

Chapter 1
Paper Instruments as Objects of Research

Abstract This chapter proposes the idea of books as objects and gives a brief introduction to paper instruments as objects of research. The paper instruments examined here are volvelles, which are moveable paper wheels from early modern textbooks that were printed together with the text of the book and installed by the reader when needed. These wheels could either work as amusement, show information, enrich a text, or help with understanding and calculating complex phenomena. The book *De sphaera* by Johannes de Sacrobosco existed as a university textbook before the "golden age" of volvelles in printed books, but began to contain volvelles in some editions during the early modern era. It thus provides a helpful place to investigate possible ideas behind the concept of volvelles for pedagogical purposes in the early modern era.

Keywords Paper instruments · Volvelles · Material culture · *Sphaera* · Johannes de Sacrobosco

> There is a long history of scientists sharing material other than words.
>
> (Baird 2004, 7)
>
> The circle has no beginning and no ending, it is unbiased, solid and unwavering in its geometric simplicity, denoting unity and eternity, totality and infinity. It represents the image of the cosmos, the cycles of the seasons, the life of man and the orbits of planets around the sun.
>
> (Helfand 2002, 13)

In order to investigate the medieval and early modern history of European science and knowledge, researchers employ many different tools and investigate countless sources, almost all written. Manuscripts and books are read and analyzed, the life and work of their authors, printers and booksellers is examined and put into historical context. In addition, scholarly research on the book itself, as well as the examination of images in historical books, contributes to our understanding of the medieval and early modern scientific world.

But there are ways of investigating the academic and scientific culture of these eras other than consulting the written word and its context. Susanna Berger and Sara Schechner, for example, recently examined an Italian painting from the seventeenth century with an emphasis on the astronomical instruments it shows (Berger and Schechner 2021).

Additionally, there is a body of scholarship concerned with historical scientific artifacts and instruments in museums and collections.[1] The knowledge concerning the manipulation and handling of these artifacts was—in most cases—transmitted orally in premodern times (De Renzi 2000, ix), and contemporary research experiences can be enriched via a material approach involving working physically with such historical artifacts. Furthermore, so called "instrument books" offer the opportunity to bring the experience of text and artifact together. These historical books often either described the use of instruments or accompanied a device (Jardine 2020, 114). In many cases the instructions contained in these works are not clear, and here techniques such as reworking experiments often help researchers to understand artifacts and afford the opportunity of gaining a deeper and "fuller understanding" of the way texts and artifacts form a functional composite (Fors et al. 2016, 89–90, Jardine 2020, 116–118).[2]

In addition to the existence of books about instruments, a special category of instruments evolved that were installed inside books and thus formed an entity with the book. These devices were installed either by the printer or the reader to enhance the text as a composition: For example, they form a moveable version of a described tool or parts of human anatomy that were treated in the text. As these devices were in nearly every case made of paper they are called "paper instruments" and belong to the category of moveable parts inside books.

Of these paper instruments, one type in particular, the paper wheels called "volvelles," will be examined here to address their function and the kind of knowledge they transferred to the books' readers. Can this combination of text, image, and instrument offer further insights into premodern scientific knowledge and thought than can be obtained from just the artifact or just the text? Volvelles are known from manuscripts in Europe from the thirteenth century (Connolly 1999; Gravelle et al. 2012; Karr Schmidt 2017, 6; Crupi 2019). The term "volvelle" descends from the Latin word *volvere* (to turn) (Gravelle et al. 2012; Oxford English Dictionary 2021) and describes an instrument that contains the function of turning a wheel, a disc or a pointer. Although these objects are usually made of paper, the term can sometimes be found describing instruments made from less ephemeral materials, such as brass or wood. The basic concept of the volvelle is still used in modern life in various contexts (Helfand 2002): Car drivers in Europe, for example, are familiar with the blue parking discs that can be used to measure the time during time-restricted

---

[1] An essential part of the community investigating historical scientific instruments is the "Scientific Instrument Commission." For more information, see https://scientific-instrument-commission.org.

[2] The historian Peter Heering, for example, re-enacted experiments with solar microscopes from the eighteenth century, giving interesting insights into not just the use of the instruments but also the written source material and the material culture of the device. For more information, see Heering (2008).

parking. Modern children's books are enriched with interactive elements such as turning wheels that give the narration a vivid experience, as in the pop-up version of *The Very Hungry Caterpillar*. And planetariums still sell paper star discs designed to help patrons explore the sky and the stars above their location in 2D. All these examples give strong indications of the general function of volvelles: One turns a wheel and something happens—information is displayed, the text enriched, or complex phenomena such as the movements of celestial bodies can be more easily understood and calculated. Especially the latter was a widely used feature in the medieval and early modern academic world, leading Nick Kanas to call them "Early Paper Astronomical Computers" (Kanas 2005). The volvelles upon which this research focuses seem to fulfill this use as well. To elucidate the background of this type of artifact, a brief overview of astronomical instruments will be given in Chapter 2, which will include the concept of "stereographic projection" and an insight into the world of hybrid books containing paper instruments such as volvelles.

The volvelles investigated in this work were found inside a book that influenced the medieval and early modern academic world for around four centuries: the *Tractatus de Sphaera* (also *De Sphaera* or *Sphaera*) by Johannes de Sacrobosco (1195–1256). The reasons for this tract's importance will be discussed in Chapter 3 of this work, which includes an overview of the knowledge evolving around and connected to the *Tractatus de Sphaera* and an insight into the research project "The Sphere," which has at its center the *Tractatus de Sphaera*.

As the corpus of the *Sphaera*-works in the European printing era contains around 359 editions, extant assembled volvelles, volvelle parts, and even wrongly assembled volvelles are numerous.[3] An overview of found volvelles and their pieces within the corpus will be given in Chapter 4.

In Chapter 5 the material culture of the four most prominent volvelles within the *Sphaera*-corpus will be examined in depth with the help of Edward McClung Fleming's Winterthur protocol and Davis Baird's concept Thing Knowledge (Fleming 1974; Baird 2004).

---

[3] In the frame of this work the terms "edition," "copy," and "book" are not used interchangeably and it is important for the research outcome to differentiate between them. The term "book" is a general term that refers to the book as an object. It includes its culture and history, and its different versions, such as copies and editions. For example, when speaking of *De Sphaera* by Johannes de Sacrobosco we are generally referring to the book, with all its history and content. The term "edition," in contrast, refers merely to a particular version of this book. For example, the very first version of the book *De Sphaera* that was printed is its first printed "edition"—but of this particular "edition" of the "book" *De Sphaera* there is not just one physical object. The print run consisted of more than one physical book, which are then called the "copies" of the first printed "edition" of the "book" *De Sphaera* by Johannes de Sacrobosco. These distinctions are necessary, on the one hand, for the historical accuracy of the occurrences, on the other, for the quantitative investigations that will be discussed in this work. The research done here relies on the data that was published under https://doi.org/10.5281/zenodo.15493638 and collected by the research project *The Sphere—Knowledge System Evolution and the Shared Scientific Identity of Europe* at the Max Planck Institute for the History of Science, Berlin. As the research corpus of the project consists of one random copy from each of the 359 editions of the book, most of the collected data is generally representative of the whole corpus of the book *De Sphaera*, but does not necessarily encompass the actual footage of assembled or incorrectly assembled volvelles shown here in Sect. 4.2.

Chapter 6 will consider the function and use of these volvelles, bringing them into context with the texts that surround them. Finally, Chapter 7 offers concluding remarks.

Bibliography

Secondary Sources

Baird, Davis. 2004. *Thing knowledge*. London: University of California Press.
Berger, Susanna, and Sara J. Schechner. 2021. Observations on Niccolò Tornioli's *The Astronomers*. *Annals of Science* 78 (4): 1–45. https://doi.org/10.1080/00033790.2021.1957149.
Connolly, Daniel K. 1999. Imagined pilgrimage in the itinerary maps of Matthew Paris. *The Art Bulletin* 81 (4): 598–622. https://doi.org/10.2307/3051336.
Crupi, Gianfranco. 2019. Volvelles of knowledge. Origin and development of an instrument of scientific imagination (13th–17th centuries). *JLIS.it* 10 (2). https://doi.org/10.4403/jlis.it-12534.
De Renzi, Silvia. 2000. *Instruments in print: Books from the Whipple collection*. Cambridge: Whipple Museum for the History of Science.
Fleming, Edward McClung. 1974. Artifact study: A proposed model. *Winterthur Portfolio* 9: 153–173.
Fors, Hjalmar, Lawrence M. Principe, and H. Otto Sibum. 2016. From the library to the laboratory and back again: Experiment as a tool for historians of science. *Ambix* 63 (2): 85–97. https://doi.org/10.1080/00026980.2016.1213009.
Gravelle, Michelle, Anah Mustapha, and Coralee Leroux. 2012. Volvelles. ArchBook. http://drc.usask.ca/projects/archbook/volvelles.php.
Heering, Peter. 2008. The enlightened microscope: Re-enactment and analysis of projections with eighteenth-century solar microscopes. *The British Journal for the History of Science* 41 (3): 345–367. https://doi.org/10.1017/S0007087408000836.
Helfand, Jessica. 2002. *Reinventing the wheel*. New York: Princeton Architectural Press.
Jardine, Boris. 2020. The book as instrument: Craft and technique in early modern practical mathematics. *BJHS Themes* 5: 111–129. https://doi.org/10.1017/bjt.2020.8.
Kanas, Nick. 2005. Volvelles! Early paper astronomical computers. *Mercury* 34 (2): 33–39.
Karr Schmidt, Suzanne. 2017. *Interactive and sculptural printmaking in the Renaissance*. Leiden: Brill. https://brill.com/view/title/33075.
Oxford English Dictionary. 2021. volvelle, n. Oxford: Oxford University Press. https://www.oed.com/view/Entry/224608?redirectedFrom=volvelles.

Open Access This chapter is licensed under the terms of the Creative Commons Attribution 4.0 International License (http://creativecommons.org/licenses/by/4.0/), which permits use, sharing, adaptation, distribution and reproduction in any medium or format, as long as you give appropriate credit to the original author(s) and the source, provide a link to the Creative Commons license and indicate if changes were made.

The images or other third party material in this chapter are included in the chapter's Creative Commons license, unless indicated otherwise in a credit line to the material. If material is not included in the chapter's Creative Commons license and your intended use is not permitted by statutory regulation or exceeds the permitted use, you will need to obtain permission directly from the copyright holder.

Chapter 2
Interactive Prints and Paper Instruments

Abstract Knowledge of the history of interactive prints, which present a moveable paper object within the printed work that can be manipulated by the reader, is crucial to understanding their function and use. Examples of interactive prints of early modern printed books such as Euclid's *Elements*, Blagrave's *The Mathematical Jewel*, or Vesalius' *Humani corporis* show how moveable parts in books were commonly used at that time. They were a helpful tool to display astronomical content in books or cheaper versions of astronomical instruments made of paper. To assist in the investigation and understanding of the use and application of volvelles—moveable paper instruments in the form of wheels—a basic outline of astronomical instruments and of the concept of stereographic projection is given in this chapter.

Keywords Material culture · Interactive prints · Paper instruments · John Blagrave · Astronomical instruments · Volvelles · Stereographic projection

Before beginning an examination of the material culture and context of medieval and early modern volvelles it is necessary to consider both the history of "interactive prints"[1] and the role of work with paper in scientific environments during the medieval and early modern era. The historians Suzanne Karr Schmidt and Boris Jardine showed that the use of paper to create specific objects offered technological innovations. It was an important tool for hands-on practice of instruments during these historical periods but could also provide entertainment in a lighter sense, when used to change images via a moveable wheel (Jardine and Grafton 1990; Karr Schmidt 2017; Jardine 2017, 2020; De Renzi 2000, 47).

Examples of interactive prints and paper instruments included not just the already mentioned volvelles, but also objects such as flaps that revealed human organs in illustrations in medical books (Fig. 2.1), pop-up geometrical figures (Fig. 2.2) or separate kits for building three-dimensional scientific instruments such as John Blagrave's (d.

---

[1] The term "interactive prints" describes printed works containing one or more moveable paper objects, that can be manipulated by the reader and thus used interactively, such as moveable flaps, tunable paper wheels or geometrical figures.

© The Author(s) 2025
A. Citron, *Spinning the Cosmos*,
SpringerBriefs in History of Science and Technology,
https://doi.org/10.1007/978-3-031-90976-4_2

1611) Jewel (Fig. 2.3). Especially paper instruments such as Blagrave's Jewel, as well as volvelles, were based on separate kits and cut-out pages that allowed the reader to build them (Fig. 2.4). Such books and interactive prints, it might be argued, required not readers but "users," and taught them to read images as interactively as they read texts (Jardine 2020, 113; Karr Schmidt 2017, 1). As Karr Schmidt states, this kind of medium "transcended the spatial limitations of the page" and created a "visual and physical contact that alters both the user and his object" (Karr Schmidt 2017, 5). These interactions with the book opened a world of mathematical, astronomical, and medical practice for those who lacked access to or knowledge of more inaccessible and valuable materials, such as brass (Jardine 2020, 121).

The extant instances of paper instruments inside books suggest that, although they can already be found in medieval manuscripts, there was something of a "golden age" for these tools shortly after the introduction of printing in Europe in the fifteenth century (Gingerich 1993, 63; Karr Schmidt 2017, 5). The new technique of printing and the lower production costs for books that came with it offered the opportunity to incorporate expensive and complex astronomical instruments, such as armillary spheres and astrolabes, into books and made them accessible to a wider audience (Kanas 2019, 264). Authors such as William Gilbert (1544–1603) encouraged their readers to cut out the diagrams in their books and paste them onto board; Blagrave offered woodcuts that were intended to be used as surrogates for their brass originals (Jardine 2020, 111, 121). Books with more complex volvelle structures appeared in the second half of the sixteenth century (Gingerich 1993, 72). Karr Schmidt has estimated that for each surviving woodcut that is known of today (around 190), at least 1000 copies of moveable books may have been lost. This implies that around 190,000 interactive prints must have been circulating in the sixteenth century (Karr Schmidt 2017, 3).

Books containing these kinds of gadgets could become "sites for mathematical practice" because they could be used and manipulated as functional instruments (Jardine 2020, 113). Additionally, it can be assumed that these paper tools served a purpose that was both pedagogical and popularizing (Gingerich 1993, 63). It may be assumed that manipulating these kinds of object provided a haptic experience that was and still is part of a fascination for instruments, which probably enriched pedagogical practice. For example, Ramón Llull (ca. 1232–ca. 1315), a Catalonian philosopher and mystic, developed a method known as *Ars magna*, for obtaining "higher knowledge of all things," of which one aspect was retrieving answers by manipulating volvelles (Lindberg 1979, 50–51; Karr 2004). Llull's concept was revolutionary because he constructed mnemonic concepts by using movement as support. In this way he made the users capable of discovering *new* knowledge by themselves, by asking questions and answering them via the manipulation of a volvelle (Karr 2004, 103–107).

Additionally, cases exist where volvelles were used as mnemonic devices in a system of learning through repetition. For example, the professor of rhetoric Jacobus Publicius worked to improve and organize the formation of memory by developing a visual alphabet where each letter was connected to an object that echoed its shape (Publicius 1482, C11–C17) (Fig. 2.5 left). This alphabet was followed by a volvelle

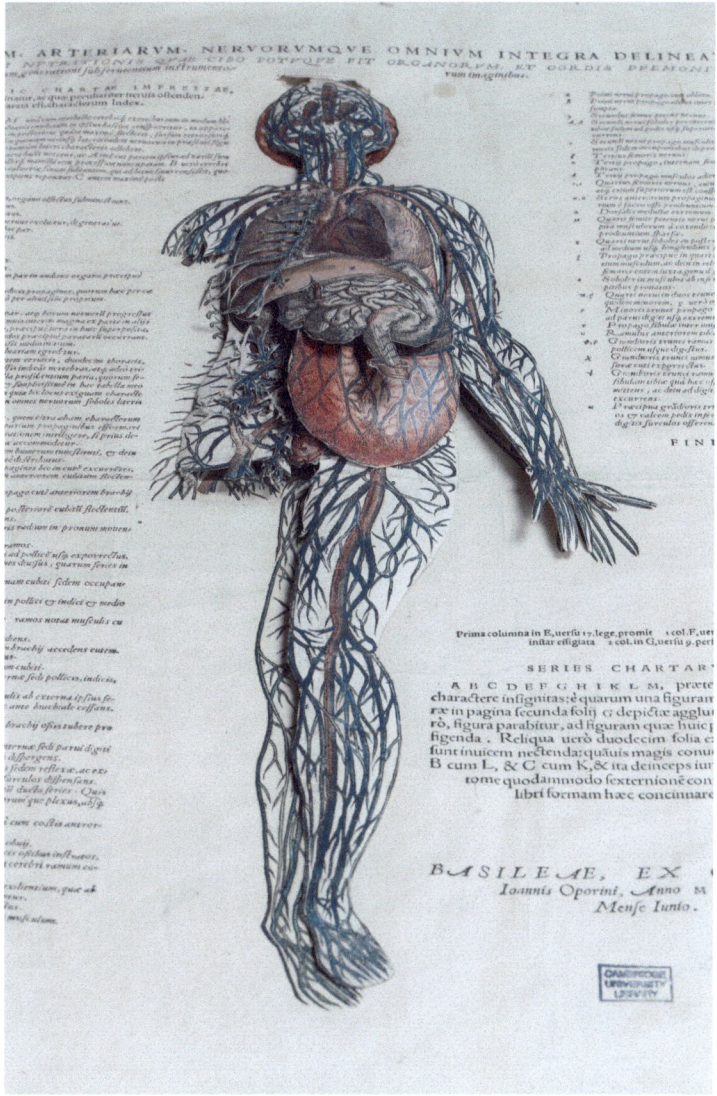

Fig. 2.1 Illustration with paper flaps showing the insides of a human body. From Vesalius (1543, 23a), Cambridge University Library

which helped to create letter combinations and patterns that were easy to remember and helped in the memorization of certain concepts (Karr 2004, 109) (Fig. 2.5 right). Publicius' idea shows vividly how illustrations and volvelles were helpful for studying basic concepts that could be hard for the mind to visualize. In particular, the study of the complex system of the celestial bodies could be supported by various instruments such as, for example, armillary spheres. Scholars and students

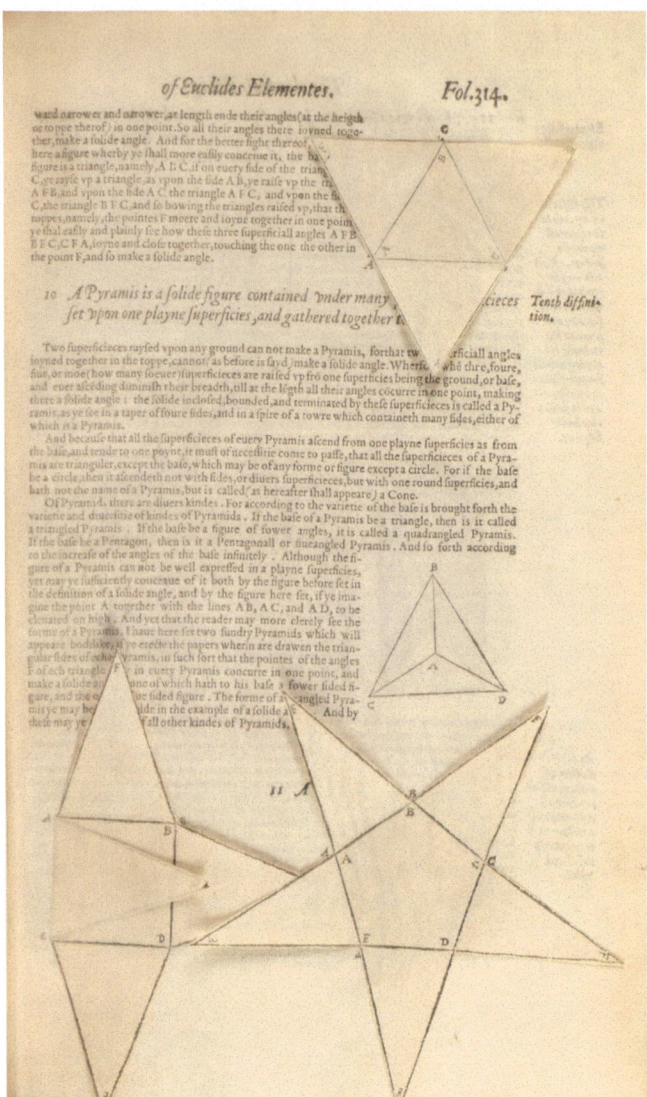

Fig. 2.2 Pop-up geometrical figures found in the English translation of Euclid's *Elements*. From Euclid (1570, fol. 314r). Courtesy of The Linda Hall Library of Science, Engineering & Technology

without access to these instruments, or wanting further knowledge and practice in manipulating more complex instruments, got the chance to participate and engage with the help of paper instruments, which filled a gap between diagrams, that could barely give all the information one needed or wanted, and instruments which were not as "easy" to access as books were.

2 Interactive Prints and Paper Instruments 11

Fig. 2.3 Assembled instrument ("Jewel") from John Blagrave's work *The Mathematical Jewel*. The instrument is cut out and pasted on board. Blagrave's jewel was intended for mathematical calculations and presented as a "universal astrolabe." From Blagrave (1585, 1v), Cambridge University Library

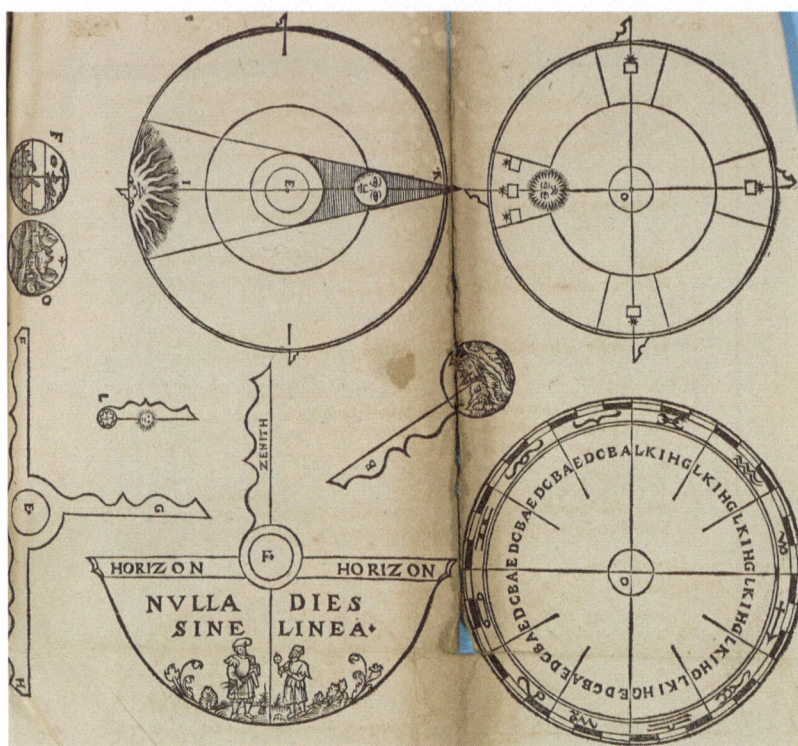

Fig. 2.4 A double page to be cut out and assembled by the reader. This separate kit contains parts of four different volvelles that could have been assembled on different pages inside the book. From Peucer (1558, foldout without number). Augsburg Staats- und Stadtbibliothek, Math 900#2, urn: nbn:de:bvb:12-bsb11267743-1

Another example of a popular and widely used pedagogical book from the medieval and early modern era containing paper tools is the thirteenth-century scholar Johannes de Sacrobosco's *Sphaera*. The first time volvelles were incorporated in the *Sphaera* was in 1538, during the already mentioned "golden age" of paper tools inside books and will be investigated later in this work.

To understand the background of the volvelles within the *Sphaera*—as well as within other historical publications—it is crucial to have a basic knowledge of the most popular historical astronomical instruments, as well as the underlying technique behind their use, called "stereographic projection." Therefore, an outline is given in the following paragraphs.

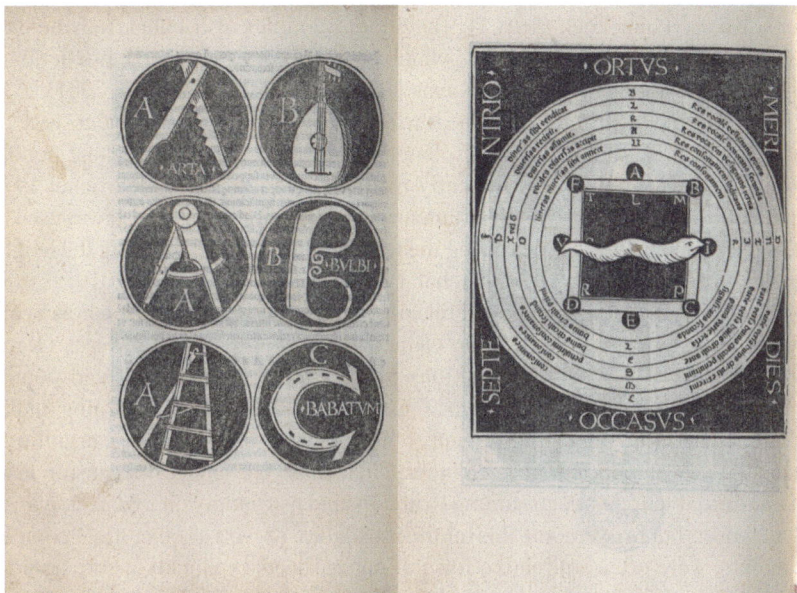

Fig. 2.5 Two pages from Jacobus Publicius's work, showing the visual alphabet and a volvelle for memorizing concepts. From Publicius (1482). Public Domain, Bibliothèque Sainte-Geneviè

2.1 Astronomical Instruments

The theoretical and observational knowledge of astronomy was already well elaborated in antiquity and made astronomy one of the first practices that evolved into a scientific discipline (Pannekoek 1961, 13; Maddison 1963; Wynter and Turner 1975, 8; Evans 1998). Navigation and the measurement and calculation of time were crucial parts of practical astronomy in the ancient world and astronomical instruments were already in use for practicing this knowledge as well as for measuring and observing the sky (Pannekoek 1961; Evans 1998). Many standard instruments developed in this period, such as quadrants, astrolabes, armillary spheres, and orreries, still exist today and are crucial to investigations of historical astronomical knowledge, tradition, and practice, and for research on scientific instruments (Wynter and Turner 1975; Dunn et al. 2018). The first secondary literature investigating scientific instruments began to appear in the second quarter of the twentieth century (Turner 1993; Jardine 2020, 115).

An example of an ancient, very simple, but also important instrument is the *gnomon*, which was probably invented in different cultures independently, as it is known not just from Babylonia and Greece but also from China before the Christian era (Pannekoek 1961, 74, 91; Evans 1998, 27). The *gnomon* was a stick that was set up vertically in a sunny place so that it cast a shadow while the sun moved across the sky. It was presumably used for time-measuring and tracking the summer and winter solstices (Pannekoek 1961, 74, 91). A more complex instrument from the same period is the Antikythera mechanism, which enabled, for example, the prediction of astronomical positions in advance (Kanas 2019, 273–274; Freeth et al. 2021).

A widely used instrument for measuring and observing the heavens was the astrolabe, which was known from Ptolemy's *Planisphaerium* but became popular in Europe in late antiquity and the early medieval era (North 1974; Michel 1976; Hartner 2021). It is a flat, circular instrument usually made of brass which was used for various purposes, including measuring the time at night and determining the heights of celestial objects (North 1974; Michel 1976; Evans 1998; Kanas 2019; Hartner 2021). Additionally, it displayed useful information on its reverse, such as scales allowing the user to find the position of the sun or the direction to Mecca (Kanas 2019, 274–278). Their various functions made astrolabes important and popular instruments that were not just of practical use but could also help in learning, understanding and memorizing the movements of the heavens, as well as the causalities of the celestial movements among each other. Their materials and craftsmanship made them unaffordable for students aiming to understand astronomy on a basic university level, however, but to overcome this hurdle astrolabes, as well as other more complex instruments, were often sold in the form of paper sheets or already incorporated in the form of paper instruments as stated above.

2.2 Stereographic Projection

An important aspect of the astrolabe, as well as of other astronomical instruments—no matter the material of construction—is the technique used to display the heaven and its objects through these instruments: the "stereographic projection." This was used extensively "during the several centuries of astrolabe use," states Snyder (1993, 291), not just for these instruments, but in the early modern era also for star maps and maps of Europe and Asia (Snyder 1993, 27). The technique describes a mapping method that helps to transfer a three-dimensional spherical surface, such as that of the world or the heavens, to a two-dimensional plane, and it is crucial knowledge when dealing with instruments that demonstrate or display the three-dimensional heaven and earth in two dimensions.[2] Stereographic projection is also an essential tool in understanding the function of volvelles that serve an astronomical purpose.

An essential part of the projection of a sphere is that some points that were separate in a three-dimensional version of the sphere are layered in the two-dimensional plane. In the example used for this chapter, these points are the north and south celestial

---

[2] Stereographic projection can also be performed with the help of mathematical calculations. For an understanding of this research a simplified geometrical understanding of the principle is functionally adequate. The mathematical knowledge and approach is not necessary, but more information on the mathematics involved and an extended geometrical approach can be found, for example, in Snyder (1993), at http://www.ams.org/publicoutreach/feature-column/fc-2014-02 (accessed 28 August 2024), in Coxeter and Greitzer (1967, 150–153) and in the context of the WolframAlpha project (https://mathworld.wolfram.com/StereographicProjection.html, accessed 28 August 2024).

2.2 Stereographic Projection

poles, which are on opposite sides in a three-dimensional shape and therefore do not touch (Fig. 2.6). Celestial parts that are considered as important for the later instrument, like tropics, the equator or the ecliptic, are projected as follows: An imaginary line is drawn from the south celestial pole to the selected part of the sky (in case of Fig. 2.7, the tropics and the equator). The celestial circle is located where this line intersects the plane of the projection in the two-dimensional map (North 1974, 99–100). A more informative visualization is perhaps the stereographic projection of the almucantars (Fig. 2.7). An almucantar is "a circle on the celestial sphere parallel to the horizon, typically one of a series that cut the meridian at equal angular separations; a parallel of altitude" (Oxford English Dictionary 2021). Two important characteristics of the stereographic projection are, first, its preservation of the angles of the three-dimensional sphere and, second, the preservation of circles, which means that the circles of the sphere will be transferred to circles in the area of the disk (Coxeter and Greitzer 1967, 152).

As stated, the technique was common during the early modern era and thus coincided with the first appearance of volvelles in editions of the *Sphaera*. But what role did volvelles perform in this treatise? Was the knowledge contained in the *Sphaera*'s text enriched or changed by the volvelles? To answer these questions, a brief introduction to the book *De Sphaera* follows.

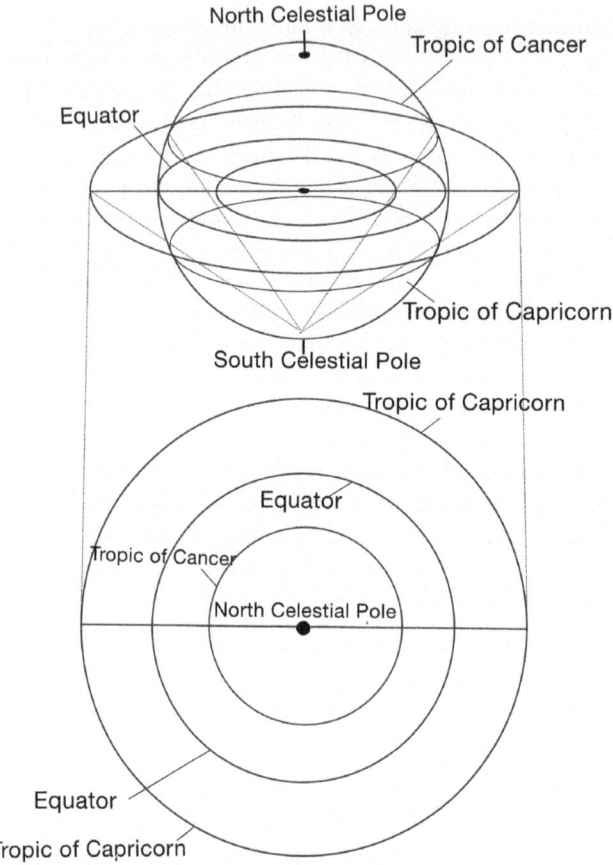

Fig. 2.6 A visual representation of stereographic projection. The upper part of the image shows the celestial sphere, including imaginary lines and two-dimensional projection. The lower part of the image shows the final two-dimensonal projection. Plot by the Author after North (1974, 99)

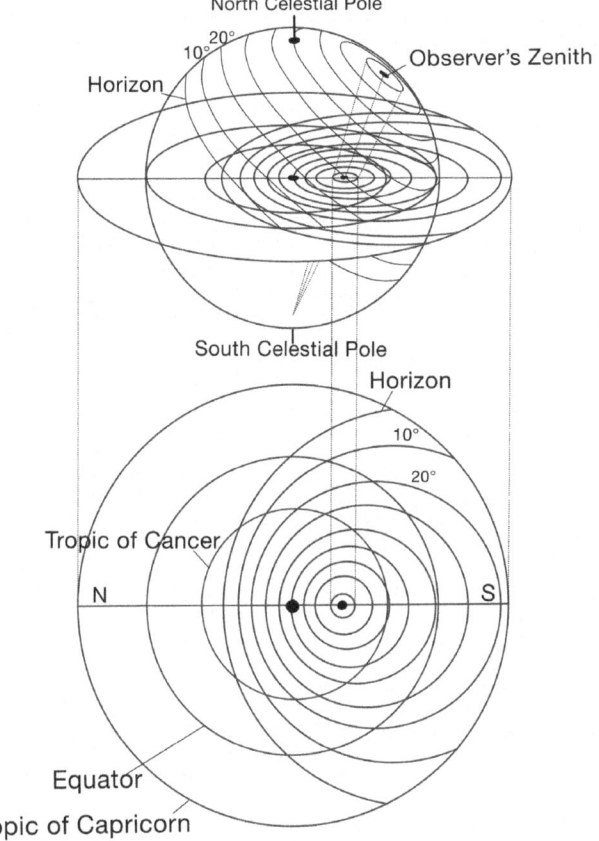

Fig. 2.7 A visual representation of stereographic projection of the almucantars. The upper part of the image shows the celestial sphere, including two-dimensional projection. The lower image shows the final two-dimensional projection. Plot by the Author after North (1974, 99)

Bibliography

Primary Sources

Blagrave, John. 1585. *The mathematical jewel*. London: Walter Venge. http://cudl.lib.cam.ac.uk/view/PR-LE-00028-00005-00007/1.

Euclid. 1570. *The Elements of geometrie of the most aunciant philosopher Euclide of Megara*. Translated by H. Billingsley. London: Iohn Daye. https://catalog.lindahall.org/discovery/delivery/01LINDAHALL_INST:LHL/12874779900059 61?lang=en&viewerServiceCode=AlmaViewer.

Peucer, Kaspar. 1558. *Elementa doctrinae de circulis coelestibus et primu motu*. Vitebergae: Crato. http://hdl.handle.net/21.11103/sphaera.100191.

Publicius, Giacomo. 1482. *Ars oratoria. Ars epistolandi. Ars memorativa*. Venice: Erhard Ratdolt. https://archive.org/details/OEXV540_P4/page/n123/mode/1up.

Vesalius, Andreas. 1543. *Andreae Vesalii suorum de humani corporis fabrica librorum epitome. Humani corporis fabrica librorum epitome.* Basilae: ex officina Ioannis Oporini. http://cudl.lib.cam.ac.uk/view/PR-CCF-00046-00036/30.

Secondary Sources

Coxeter, H.S.M., and Samuel L. Greitzer. 1967. *Geometry revisited.* Mathematical Association of America. New York: Random House. https://www.cambridge.org/core/books/geometry-revisited/C220BC3BF7C2634DBF4F3BB7291B6E19.

De Renzi, Silvia. 2000. *Instruments in print: Books from the Whipple collection.* Cambridge: Whipple Museum for the History of Science.

Dunn, Richard, Silke Ackermann, and Giorgio Strano. 2018. *Heaven and earth united: Instruments in astrological contexts.* Leiden: Brill. https://brill.com/view/title/36320.

Evans, James. 1998. *The history and practice of ancient astronomy.* New York/Oxford: Oxford University Press.

Freeth, Tony, David Higgon, Aris Dacanalis, Lindsay MacDonald, Myrto Georgakopoulou, and Adam Wojcik. 2021. A model of the cosmos in the ancient Greek Antikythera Mechanism. *Scientific Reports* 11 (1): 5821. https://doi.org/10.1038/s41598-021-84310-w.

Gingerich, Owen. 1993. Astronomical paper instruments with moving parts. In *Making instruments count. Essays on historical scientific instruments presented to Gerard L'Estrange Turner*, ed. Robert G.W. Anderson, Jim Bennett, and William F. Ryan, 63–74. Aldershot: Variorum.

Hartner, Michael. 2021. Asturlāb. In *Encyclopaedia of Islam*, 2nd ed., 722–728. Leiden: Brill.

Jardine, Boris. 2017. State of the field: Paper tools. *Studies in History and Philosophy of Science Part A* 64: 53–63. https://doi.org/10.1016/j.shpsa.2017.07.004.

Jardine, Boris. 2020. The book as instrument: Craft and technique in early modern practical mathematics. *BJHS Themes* 5: 111–129. https://doi.org/10.1017/bjt.2020.8.

Jardine, Lisa, and Anthony Grafton. 1990. 'Studied for action': How Gabriel Harvey read his livy. *Past & Present* 129: 30–78.

Kanas, Nick. 2019. *Star maps.* Cham: Springer.

Karr, Suzanne. 2004. Constructions both sacred and profane: Serpents, angels, and pointing fingers in Renaissance books with moving parts. *The Yale University Library Gazette* 78 (3/4): 101–127.

Karr Schmidt, Suzanne. 2017. *Interactive and sculptural printmaking in the Renaissance.* Leiden: Brill. https://brill.com/view/title/33075.

Lindberg, Sten G. 1979. Mobiles in books. Volvelles, inserts, pyramids, divinations, and children's games. *The Private Library* 2 (2): 41–82.

Maddison, Francis. 1963. Early astronomical and mathematical instruments: A brief survey of sources and modern studies. *History of Science* 2 (1): 17–50. https://doi.org/10.1177/007327536300200102.

Michel, Henri. 1976. *Treatise on the astrolabe.* Paris: Librairie Alain Brieux.

North, John. 1974. The astrolabe. *Scientific American* 230 (1): 96–106.

Oxford English Dictionary. 2021. almucantar, n. Oxford: Oxford University Press. https://www.oed.com/view/Entry/5616?redirectedFrom=almucantar.

Pannekoek, Anton. 1961. *A history of astronomy.* New York: Interscience Publishers.

Snyder, John P. 1993. *Flattening the earth.* Chicago, IL: University of Chicago Press. https://urn.nb.no/URN:NBN:no-nb_digibok_2009081201045.

Turner, Anthony. 1993. Interpreting the history of scientific instruments. In *Making instruments count. Essays on historical scientific instruments presented to Gerard L'Estrange Turner*, ed. Robert G.W. Anderson, Jim Bennett, and William F. Ryan, 17–26. Ashgate: Variorum.

Wynter, Harriet, and Anthony Turner. 1975. *Scientific instruments.* New York: Charles Scribner's Sons.

Open Access This chapter is licensed under the terms of the Creative Commons Attribution 4.0 International License (http://creativecommons.org/licenses/by/4.0/), which permits use, sharing, adaptation, distribution and reproduction in any medium or format, as long as you give appropriate credit to the original author(s) and the source, provide a link to the Creative Commons license and indicate if changes were made.

The images or other third party material in this chapter are included in the chapter's Creative Commons license, unless indicated otherwise in a credit line to the material. If material is not included in the chapter's Creative Commons license and your intended use is not permitted by statutory regulation or exceeds the permitted use, you will need to obtain permission directly from the copyright holder.

Chapter 3
The *Sphaera* of Johannes de Sacrobosco in the Early Modern Period

Abstract Shortly after Johannes de Sacrobosco published his astronomical treatise *De sphaera* in the thirteenth century it began a process of transformation and a commentary tradition over more than 350 editions within the next 400 years. This commentary tradition helped in shaping the development of scientific enquiry in Europe by circulating continuously adapted and transformed ideas and knowledge. In due course, it became one of the most influential textbooks of the early modern university. This chapter describes the knowledge tradition in and around the *Sphaera*, as well as its contents, and the research project *The Sphere* at the Max Planck Institute for the History of Science. This project investigates a corpus of 359 *Sphaera* editions and explores the textual and visual features of the editions, while developing hypotheses on the knowledge disseminated through the textual and visual levels of the book. For this, several computational tools were developed within the framework of the project that were used for the investigation of the volvelles contained in the corpus.

Keywords *Sphaera* · Johannes de Sacrobosco · Early modern period · History of science · History of knowledge · Machine learning · Digital history · Database

In the thirteenth century a scholar teaching in Paris, Johannes de Sacrobosco, wrote a treatise which would become one of the most used textbooks in astronomy and cosmography from the thirteenth to the seventeenth centuries (Thorndike 1949; Grant 1996, 20; Hamel 2014). About Sacrobosco, amazingly little is known. Probably he was born towards the end of the twelfth century in England or Ireland. He could have studied in Oxford and apparently became an Augustinian monk (Hamel 2014, 9). During his time at the university in Paris, where his presence was attested from June 1221, he wrote one of the most successful scientific books of all times: *Tractatus de Sphaera* (Hamel 2014, 7).

What made Sacrobosco's treatise so popular and important, not just within the context of the early modern university? First, it is necessary to know that Sacrobosco's lifetime and the time of the *Sphaera*'s creation was a significant period for astronomy: In 1175 Gerard of Cremona (1114–1187) translated the *Almagest*, written by Ptolemy

(AD 100–170), from Arabic into Latin. As the *Almagest* promoted a geocentric model that formed the paradigm for the astronomical discipline for hundreds of years it was already one of the most important textbooks for astronomers before the *Sphaera* was even written (*Encyclopedia Britannica* 2002). Subsequent to Gerard of Cremona's translation, a continuous wave of translations of Arabic works into Latin emerged that "brought breadth of knowledge and a more comprehensive view of the world" to Europe (Delisle et al. 2012, 110).

On this fertile ground fell Sacrobosco's *Tractatus de Sphaera*. Despite its elementary character, the treatise contained the entire matter of astronomy at that time, influenced by some of the translations described above, as well as the works of Al-Farghānī (Abū al-ʿAbbās Aḥmad ibn Muḥammad ibn Kathīr al-Farghānī, 800–870) and Euclid's (ca. third century BCE) *Elements* (Thorndike 1949, 1, 14–16; Hamel 2014, 10–11). As already mentioned, it soon became one of the most used textbooks in astronomy of that time. The traces of Sacrobosco's work can be found in the medieval period through manuscripts as well as in the print works of the early modern period.

Of course, Sacrobosco was not the only author compiling a work of this character at that time: Robert Grosseteste (1170–1253), John Peckham (1230–1292), and Campanus of Novara (1220–1296) wrote rival texts against which Sacrobosco's *Sphaera* prevailed (Thorndike 1949, 23; Valleriani 2017, 427).

Sacrobosco's work seemed to meet the concerns of its time on an intellectual and institutional level, but its success is primarily grounded in the medieval university system and its requirements (Hamel 2014, 10; Valleriani 2017, 427). Students in the medieval period had to study the so called "seven liberal arts" (*septem artes liberals*), which required four years of work in dialectic, grammar, rhetoric, arithmetic, geometry, music, and astronomy (Grant 1996, 20; Hamel 2014, 10). As an introductory reading the *Sphaera* became the standard work for studying astronomy within this framework. Over time the book became aimed more and more strongly at the student market; in some copies, students were even named as recipients (Hamel 2014, 11; Valleriani et al. 2019, 56). The *Sphaera* was translated early into various vernacular languages: first into Old Islandic in 1250, followed by, among others, German, Italian, and French (Hamel 2014, 11). Today more than 300 handwritten manuscripts of the *Sphaera* are extant, in addition to 359 printed editions—all part of the knowledge tradition of the *Sphaera* (Hamel 2014, 11; The Sphere 2021).[1] Owen Gingerich has estimated that a total of more than 200,000 copies of the *Sphaera* must have been printed (Gingerich 1990, 190), a number he based on the 200 print editions that were known in 1990. Taking into consideration the now in total known 359 print editions discovered by The Sphere research project, there were probably more than 350,000 printed copies of the *Sphaera*.

---

[1] The project *The Sphere—Knowledge System Evolution and the Shared Scientific Identity of Europe* at the Max Planck Institute for the History of Science investigates the knowledge tradition evolving around Johannes de Sacrobosco's book *The Sphere*. More about the project can be found in Sect. 3.2.

3.1 Knowledge Within the *Sphaera*

To understand the knowledge tradition of the *Sphaera*, it is necessary to first shed light on the its contents and the topics discussed in it. Like previous scholars, I find Sacrobosco's own words to be most helpful in describing the principles of his four chapters:

> The treatise on the sphere we divide into four chapters, telling, first, what a sphere is, what its center is, what the axis of a sphere is, what the pole of the world is, how many spheres there are, and what the shape of the world is. In the second we give information concerning the circles of which this material sphere is composed and that super celestial one, of which this is the image, is understood to be composed. In the third we talk about the rising and settings of the signs, and the diversity of days and nights which happens to those inhabiting diverse localities, and the division into climes. In the fourth chapter the matter concerns the circles and motions of the planets, and the causes of eclipses [of the sun and the moon].[2]

These topics were discussed across a total of about forty pages, this limited treatment already showing the elementary character of the work. Nevertheless, students and readers gained not just a basic understanding of the concepts of heaven at that time but were equipped with knowledge on the cosmos, the elements, meteorology, eclipses, and more.

Reading a text in medieval and early modern times was equivalent to "working with the book" or "using it."[3] It was quite a common practice to add handwritten notes to books or manuscripts while reading to gain a better understanding or memorization of the content.[4] These notes, called glosses, were a typical method for adding knowledge to already existing works (Meyers Konversationslexikon 1885–1892; Brockhaus Konversationslexikon 1894–1896; Hannebutt-Benz 1989). The practice developed in the historical schools of jurisprudence in Europe, though examples already existed in first-century CE German-language manuscripts (Brockhaus Konversationslexikon 1894, 1896), and was continued by adding printed glosses to texts after the introduction of the printing technique in Europe. An important aspect of the tradition around the *Sphaera* is this commentary tradition, which evolved around the original treatise shortly after its first publication. The knowledge that was codified in the *Sphaera* was not just translated into other languages, thus being expanded and transmitted to other scientific communities, but was also complemented by comments, changed text parts, or glosses, especially in the printing era, where the glosses were

---

[2] Sacrobosco (1472): "*Tractatum de spera quatuor capitulis distinguimus dicentes, primo quid sit spaera, quod eius centrum, quid axis spaerae, quid sit polus mundi, quot sind spaerae et quae sit forma mundi. In secundo de circulis ex quibus haec spaera materialis componitur et illa supercoelestis quae per istam ymaginatur componi intelligitur. In tercio de ortu et occasu sigonorum de diversitate noctium et dierum et de devisione climatum. In quarto de circulis et motu planetarum et de causis eclipsium solis et lunae.*" Translated by Thorndike (1949, 76).

[3] For more information on the practice of reading through different periods see Hannebutt-Benz (1989).

[4] Lots of the collected books in the corpus of the project *The Sphere* contain handwritten notes. For more information on the practice see Sherman (2009), who investigated in his work the tradition of how books were used and read in early modern England.

printed together with the original text. The existing corpus of printed *Sphaera* editions between 1472 and 1650 consists of more than 150 works that contain glosses and comments in different forms: *glossae interlineares* (interlinear glosses), which were added between the lines of the original text, and could also merge with the text and build a new "apparatus" (Fig. 3.1); or *glossae marginales* (marginal glosses), which were added to the margin of the book (Fig. 3.2) (Meyers Konversationslexikon 1885, 1892; Schneider 2016). *Glossae marginales* surrounding the original text were termed *Schachtelglosse* (box gloss); those that touched the text on two or three sides were called *Klammerglosse* (bracket gloss) (Fig. 3.3) (Hannebutt-Benz 1989; Schneider 2016).

Through this technique of constantly explaining and adding new or complementing knowledge to the topics discussed within the *Sphaera*, the knowledge base of the book transformed in various ways over the centuries. Of the extant printed copies of *Sphaera*, nearly half contain commented material that extended the already codified knowledge inside the book. Print masters that were aware of the market and the material added the knowledge of authors such as Christoph Clavius (1538–1612), Oronce Finé (1494–1555), and Johannes Regiomontanus (1436–1476), as well as new subjects to the works (Valleriani and Ottone 2022).[5] In addition, knowledge from authors whose textbooks were considered too expensive or technical was added to the *Sphaera* (Gingerich 1990, 191).

Another means of reshaping the *Sphaera* was the *Quaestio*: a question format in which questions were posed and directly answered using the contents of the original *Sphaera* text. This format evolved from lectures and survived in its written form (Grant 1996, 23–24).

With the help of the commentary tradition, the *Sphaera* underwent a transformational process that supplemented the original knowledge contained within. The subject areas touched on by the *Sphaera* over the centuries included, for example, mathematical astronomy, calendrical calculations, the use of and instructions for constructing astronomical instruments, nautical astronomy and geography, cartography, meteorology, arithmetic, geometry, judiciary astrology, literature, practical optics, and mechanics (Valleriani 2017, 431).[6]

In addition to the commentary tradition, the original as well as the commented and annotated *Sphaera* works were bound together with other works to create a compilation of texts. The introductory character of the *Sphaera*, together with more complex works, formed an entire course on the astronomical discipline at that time (Hamel 2014, 14). Works that were often bound together with the *Sphaera* included,

---

[5] For the roles of Christoph Clavius, Oronce Finé, and Johannes Regiomontanus in the frame of the Sphaera see http://hdl.handle.net/21.11103/sphaera.100732 (Clavius); http://hdl.handle.net/21.11103/sphaera.100760 (Finé); http://hdl.handle.net/21.11103/sphaera.100945 (Regiomontanus).

[6] Matteo Valleriani investigated the knowledge contained in the *Sphaera* with the help of the network theory in Valleriani (2017).

3.1 Knowledge Within the *Sphaera*

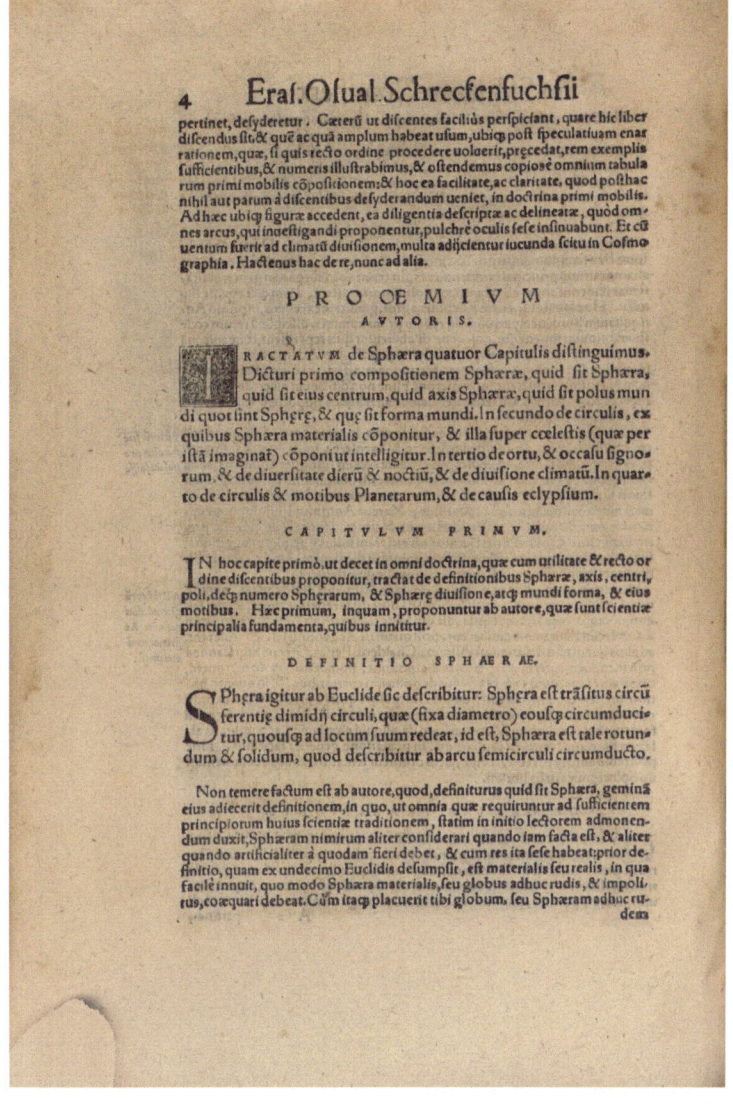

Fig. 3.1 *Glossae interlineares*, which are separated with a different typeface from the original text. From Schreckenfuchs et al. (1569, 4). BSB München, 2 Astr.u. 50, urn:nbn:de:bvb:12-bsb10141204-0

for example, the *Theorica planetarum* by Gerard of Cremona and the *Theorica novae planetarum* by Georg von Peuerbach (1423–1461).[7]

---

[7] For the roles of Gerhard of Cremona and Georg von Peuerbach in the frame of the *Sphaera* see http://hdl.handle.net/21.11103/sphaera.100773 (Cremona) and http://hdl.handle.net/21.11103/sphaera.100965 (Peuerbach).

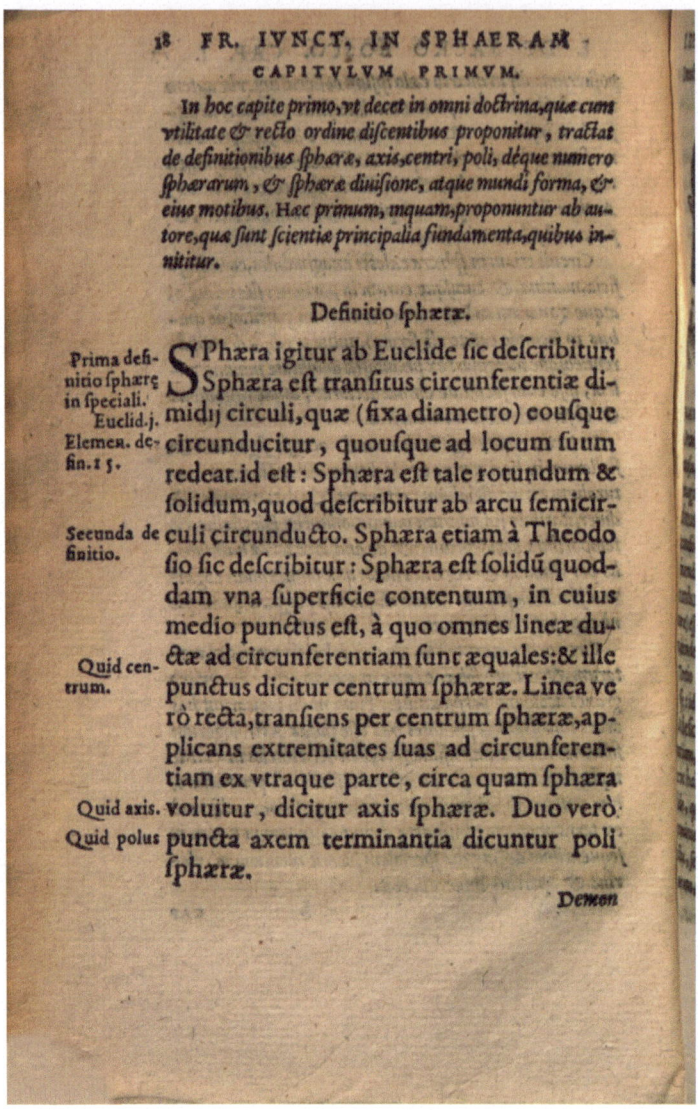

Fig. 3.2 *Glossae marginales* with the beginning of the original *Sphaera* text. From Giuntini and Sacrobosco (1577, 18). Augsburg Staat- und Stadtbibliothek, Math 382, urn:nbn:de:bvb:12-bsb112 67512-9

The way in which the knowledge contained in the *Sphaera* changed over time was a consequence of the "intellectual needs for a new knowledge system in astronomy and cosmology" in the early modern era, which was probably connected to the rising mobility of people seen at that time (Valleriani 2017, 434). Its title became a term that united a particular collection of astronomical texts, as outlined above, and was a

3.1 Knowledge Within the *Sphaera* 27

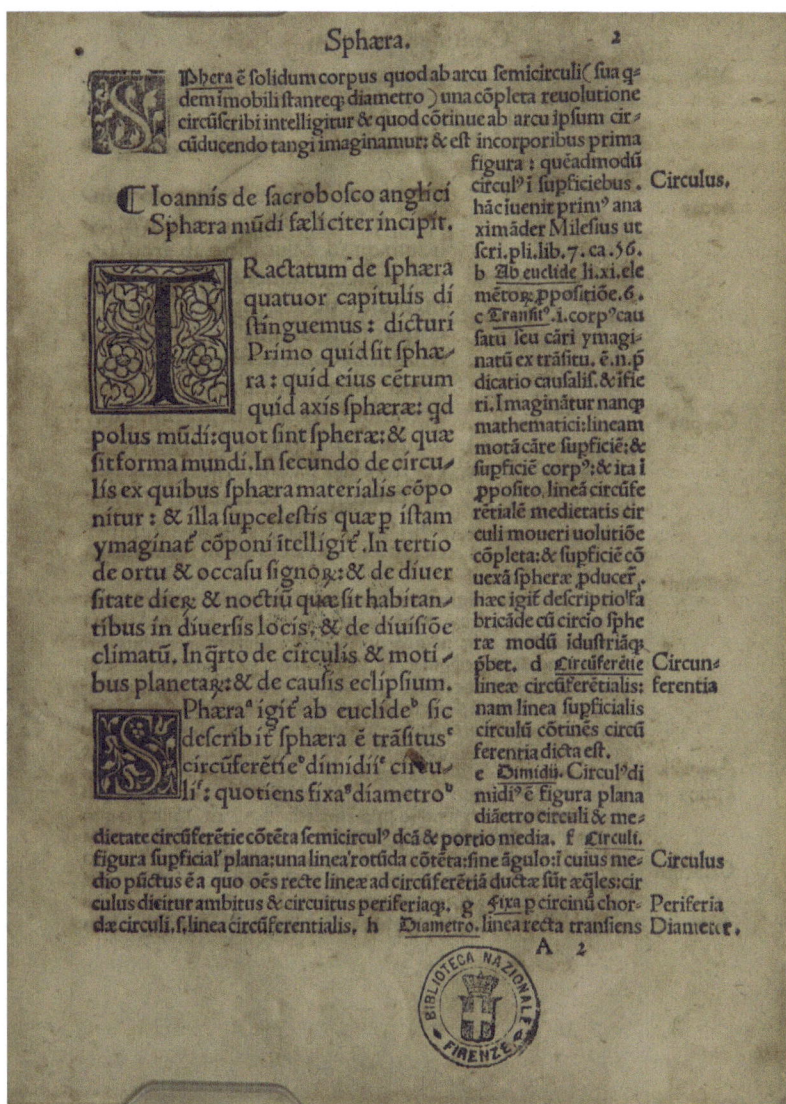

Fig. 3.3 *Klammerglosse* and *glossae marginales* inside a *Sphaera* work. From Ferrariis (1500, fol. 2r). Public Domain Mark, archive.org

"stable label for the continued accumulation and mutation of knowledge" (Valleriani 2017, 15). The texts under this label developed through the addition of commentaries and supplemental knowledge, as well as new topics, which formed the commentary tradition around the *Sphaera*. The codification of practical knowledge within the book and others in this tradition contributed to and helped to shape the scientific identity of Europe (Valleriani 2017, 429, 460).[8]

3.2 The Sphere—A Research Project

The knowledge tradition around Johannes de Sacrobosco's *De Sphaera* is investigated in a special research project at the Max Planck Institute for the History of Science (MPIWG) in Berlin under the aegis of Matteo Valleriani called *The Sphere: Knowledge System Evolution and the Shared Scientific Identity of Europe*.[9] The goal of the project was to trace the knowledge-transformation process the *Sphaera* underwent with the beginning of the printing era in Europe (see Chapter 3).

To draw representative conclusions about this tradition and the works included within it, only printed editions between 1472 and 1650 are investigated in the frame of the project, as the first *Sphaera* was printed in 1472 in Ferrara and its relevance to university teaching declined around 1650 (Valleriani et al. 2019, 52). The first steps of the research project were to create a corpus by collecting the extant copies of printed *Sphaera* treatises and to create an elaborate taxonomy for the different book types that had evolved through the centuries. The five book types that could be identified were:

1. Original treatise (16 books): Sacrobosco's treatise printed as a standalone monograph, in the original Latin or in translation.
2. Annotated original (48 books): Sacrobosco's treatise commented on or annotated by a different author, keeping the original text as its basis.
3. Compilation of texts (44 books): Sacrobosco's treatise included among other treatises by different authors. Short paratexts by different authors or compilations of several texts by one author are not subsumed under this category.
4. Compilation of texts and annotated original (124 books): Sacrobosco's text featured as the basis for a commentary or annotation and included among other treatises by different authors.
5. Adaption (127 books): Treatises and works that significantly resemble Sacrobosco's treatise in terms of content and structure but do not include the original text. Although implying a certain vagueness, this book type is relevant here in order to represent Sacrobosco's treatise as an exemplar of the period's cosmological writings. These are most often simply called *De sphaera* and are structured in four parts, like the original treatise. This book type also includes treatises

---

[8] For more detailed information on the practical knowledge within the Sphere and how the exact codification and transformation of it influenced the European past see Valleriani (2017).

[9] For more information on the project *The Sphere and access to the database*, see https://sphaera.mpiwg-berlin.mpg.de.

3.2 The Sphere—A Research Project

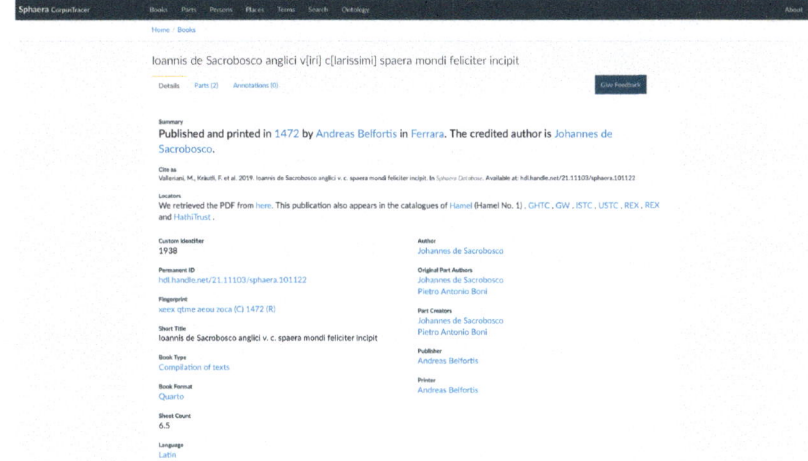

Fig. 3.4 Dataset of the first printed *Sphaera* edition from Ferrara. http://hdl.handle.net/21.11103/sphaera.101122

called *Quaestiones*, which deal with the topics of Sacrobosco's treatise in the form of questions and answers.[10]

To handle the corpus data and create a comprehensive environment for continuously investigating the source material a database was set up which is open and free to the public. To maintain the readability and accessibility into the future the database works with standards such as the CIDOC Conceptual Reference Model (CRM), which was adjusted according to the needs of the project.[11]

Within the database each book has its own record, which interacts with different possibilities inside the database and has the following features that can serve different research purposes:

- the **metadata** (including title, author(s), printer(s), publisher(s), place of publication, year(s) of publication, fingerprint) of every book is saved and stored in a standardized manner[12] (Fig. 3.4).

---

[10] These book types and their explanation are directly taken from the project's website: "The Corpus," https://sphaera.mpiwg-berlin.mpg.de/database/, accessed 28 August 2024.

[11] CIDOC-CRM is "a theoretical and practical tool for information integration in the field of cultural heritage. It can help researchers, administrators and the public explore complex questions with regards to our past across diverse and dispersed datasets. The CIDOC CRM achieves this by providing definitions and a formal structure for describing the implicit and explicit concepts and relationships used in cultural heritage documentation and of general interest for the querying and exploration of such data." http://www.cidoc-crm.org, accessed 28 August 2024. How the tool was specifically adjusted is described in Kräutli and Valleriani (2018).

[12] A fingerprint helps to clearly identify a certain book with a special technique. For the workflow involved in creating a fingerprint and background information see Beyer (2019).

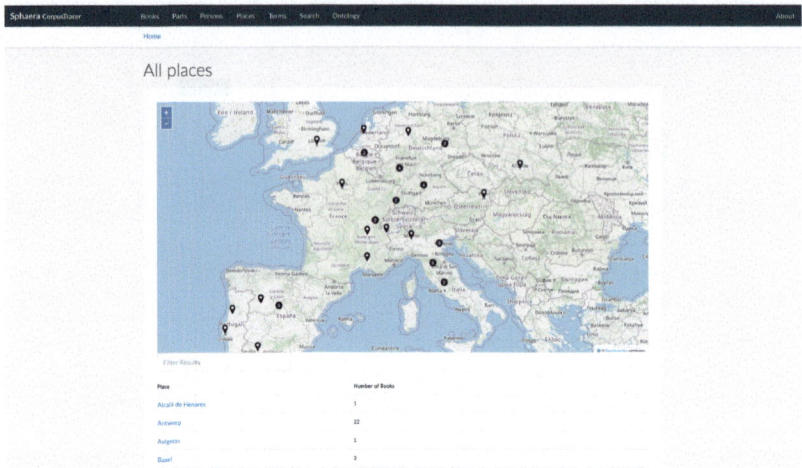

Fig. 3.5 Map showing the European printing houses that printed the *Sphaera* between 1472 and 1650. http://db.sphaera.mpiwg-berlin.mpg.de/resource/:listPlaces

- every book record is linked to different **locators** that provide digital facsimiles or library records, of which at least one was inspected before the publication of the database.[13]
- **places of publication** have been translated to their modern names and can be visualized on a map (Fig. 3.5).
- the **authors**' names are linked to datasets that give overviews of their work, in addition to short biographies that are linked to WikiData or other databases; **printers**' and **publishers**' information is treated in the same way.
- a "**Parts**" tab provides information on the individual text parts of a book, including its table of contents and page numbers.
- an "**Annotations**" tab can provide further information on the work that was inspected for the record.[14]

During the research project different aspects of the tradition around the *Sphaera* were investigated, often embedded in digital tools, databases or interfaces that were tailored to the need of the research question by the Digital Humanities support of the project that also provided the database. A variety of publications from the project investigated, for example, the epistemic communities in the *Sphaera* corpus, the special role of the authors of different *Sphaera* works and translations, and the role of the printers (Valleriani et al. 2019; Valleriani 2020; Valleriani and Ottone 2022).

---

[13] The collected copies, which now each represent an edition of a *Sphaera*, were picked according to practical reasons, such as availability. For each copy other copies exist around the world, which can be identified by the *Locators*.

[14] The website of the project gives a similar overview on its features on the page "Database" (https://sphaera.mpiwg-berlin.mpg.de/database/, accessed 28 August 2024).

Another topic handled during the research project was the images found inside the works. With the help of the database (a part not publicly accessible) and a machine learning algorithm, the images inside the 359 works were captured. More than 20,000 images were identified and sorted by aspects such as type of image, place, and so on.[15] In the final phase of the project the images were categorized and analyzed by researchers with the aim of gaining a better and more profound understanding of their role within the knowledge tradition of the *Sphaera* and the early modern period more generally. A tool was programmed by the project's data specialists for clustering images by distinct patterns and characteristics. During the process of analyzing and categorizing the images the role of volvelles within the corpus was repeatedly discussed. Do they (at least partly) count as images, or are they instruments? How do we clearly recognize parts of volvelles that were not assembled? And what do the volvelles and their material culture contribute to the *Sphaera* tradition as a whole? The conclusion reached as a result of these debates was that volvelles have to be treated differently because of their special characteristics, the haptic role they play, and the fragmented source material that requires identification. Therefore, with the help of the image clustering tool, all volvelle parts were placed in one group that could be definitely identified as parts of or complete volvelles, to allow the ongoing analysis of the source material. The method, process, and results of this research will be discussed in the following chapters.

Bibliography

Digital Repositories

Sphaera Corpus*Tracer*   Max Planck Institute for the History of Science. https://db.sphaera.mpiwg-berlin.mpg.de/resource/Start.

Primary Sources

Ferrariis, Georgius de. 1500. *Figura sphere: cum glosis Georgii de Monteferrato artium et medicine doctoris*. Venice: Giovanni Battista Sessa I. http://hdl.handle.net/21.11103/sphaera.100275.
Giuntini, Francesco, and Johannes de Sacrobosco. 1577. *Fr. Iunctini Florentini, sacrae theologiae doctoris, commentaria in sphaeram Ioannis de Sacro Bosco accuratissima. Omnia iudicio S.R. Ecclesiae submissa sunto*. Lyon: Philippe Tinghi. http://hdl.handle.net/21.11103/sphaera.100921.
Sacrobosco, Johannes de. [1472]. *Tractatum de spera*. Venice: Florentinus de Argentina.
Schreckenfuchs, Oswald, Johannes Regiomontanus, and Johannes de Sacrobosco. 1569. *Erasmi Osvvaldi Schreckenfuchsii commentaria, in sphaeram Ioannis de Sacrobusto, accuratissima,*

---

[15] The images from the project are visualized and can be accessed here: https://sphaera.mpiwg-berlin.mpg.de/coins/, accessed 28 August 2024.

quibus non solum ea quae in autoris contextu sunt, sed alia etiam ad sphaericam doctrinam necessaria, explicantur: Tabularum atque constructio, ex suis principiis per demonstrationum seriem clarem dilucide atque docetur. His adiecti sunt eiusdem autoris canones, quibus usus tabularum, quae operi ex libro directionum Ioannis Regiomontani, passim inseruntur, ad pulcherrimas inquisitiones sstronomicas, luculentissime continetur. Reliqua ad consummatam doctrinam hanc pertinentia, ex illum primo mobili, eadem forma edito, petes. Basel: Heinrich Petri. http://hdl.handle.net/21.11103/sphaera.101080.

Secondary Sources

Beyer, Victoria. 2019. *How to generate a fingerprint?* Preprint. Berlin: Max Planck Institute for the History of Science. https://www.mpiwg-berlin.mpg.de/preprint/how-generate-fingerprint.

Brockhaus Konversationslexikon. 1894–1896. Glosse. In *8. Band: Gilde—Held*. Berlin/Wien: Firma Brockhaus in Leipzig.

Delisle, Jean, Judith Woodsworth, and Judith Woodsworth. 2012. *Translators through history: Revised edition*. Philadelphia, PA/Amsterdam: John Benjamins. http://ebookcentral.proquest.com/lib/mpiwissberlin-ebooks/detail.action?docID=949206.

Encyclopedia Britannica. 2002. Almagest. https://www.britannica.com/topic/Almagest.

Gingerich, Owen. 1990. Five centuries of astronomical textbooks and their role in teaching. *International Astronomical Union Colloquium* 105: 189–196. https://doi.org/10.1017/S0252921100086711.

Grant, Edward. 1996. *Plantes, stars and orbs. The medieval cosmos, 1200–1687*. Cambridge: Cambridge University Press.

Hannebutt-Benz, Eva-Maria. 1989. *Die Kunst des Lesens. Lesemöbel und Leseverhalten vom Mittelalter bis zur Gegenwart*. Frankfurt am Main: Museum für Kunsthandwerk.

Hamel, Jürgen. 2014. *Studien zur "Sphaera" des Johannes de Sacrobosco, acta historica astronomiae*. Leipzig: AVA, Akademische Verlagsanstalt.

Kräutli, Florian, and Matteo Valleriani. 2018. CorpusTracer: A CIDOC database for tracing knowledge networks. *Digital Scholarship in the Humanities* 33 (2). https://doi.org/10.1093/llc/fqx047.

Meyers Konversationslexikon. 1885–1892. Glosse. In *7. Band: Gehirn—Hainichen*. Leipzig/Wien: Verlag des Bibliographischen Instituts.

Schneider, Daniel. 2016. *Besonderheiten mittelalterlicher Texte*. Trier: Virtuelles Museum DH. https://dhmuseum.uni-trier.de/node/378.

Sherman, William H. 2009. *Used books. Marking readers in Renaissance England*. Philadelphia: University of Pennsylvania Press.

The Sphere. 2021. *Sphaera Corpus*Tracer. Max Planck Institute for the History of Science. https://sphaera.mpiwg-berlin.mpg.de/database/.

Thorndike, Lynn. 1949. *The sphere of Sacrobosco and its commentators*. Chicago: The University of Chicago Press.

Valleriani, Matteo. 2017. *The structures of practical knowledge*. Ed. Matteo Valleriani. Dordrecht: Springer.

Valleriani, Matteo. 2020. *De sphaera of Johannes de Sacrobosco in the early modern period: The authors of the commentaries*. Cham: Springer.

Valleriani, Matteo, and Andrea Ottone. 2022. *Publishing Sacrobosco's De sphaera in early modern Europe: Modes of material and scientific exchange*. Cham: Springer.

Valleriani, Matteo, Florian Kräutli, Maryam Zamani, Alejandro Tejedor, Christoph Sander, Malte Vogl, Sabine Bertram, Gesa Funke, and Holger Kantz. 2019. The emergence of epistemic communities in the Sphaera corpus. *Journal of Historical Network Research* 3 (1): 50–91. https://doi.org/10.25517/jhnr.v3i1.63.

Open Access This chapter is licensed under the terms of the Creative Commons Attribution 4.0 International License (http://creativecommons.org/licenses/by/4.0/), which permits use, sharing, adaptation, distribution and reproduction in any medium or format, as long as you give appropriate credit to the original author(s) and the source, provide a link to the Creative Commons license and indicate if changes were made.

The images or other third party material in this chapter are included in the chapter's Creative Commons license, unless indicated otherwise in a credit line to the material. If material is not included in the chapter's Creative Commons license and your intended use is not permitted by statutory regulation or exceeds the permitted use, you will need to obtain permission directly from the copyright holder.

Chapter 4
Volvelles in the *Sphaera* Corpus

Abstract The existing editions of the treatise *De sphaera* by Johannes de Sacrobosco testify to a centuries-long commentary tradition transmitting and circulating astronomical knowledge throughout Europe during the early modern period. Since the treatise continuously changed and transformed so did the visual appearance of the work, in terms of not just the images that were added but also the paper instruments, so called volvelles, that began to appear in books in the sixteenth century. The first appeared in *De sphaera* in an edition by the printer Joseph Klug of Wittenberg. Over the course of the sixteenth century, several different volvelles were designed and printed, forming three groups, which I called the Wittenberg group, the Seville group, and the Leiden group. This chapter describes how the groups were formed, which volvelles they contain, and how the data for their investigation was collected with the help of tools such as *CorDeep* and the *Sphaera Infrastructure Tool*, which were developed within the research project The Sphere. In addition, a brief insight into the situation of the *Privilegia Impressoria* in the context of the volvelles is given in order to open the discussion for desiderata in this direction.

Keywords Sphaera · Volvelles · Paper instruments · Early modern period · History of science · History of knowledge · Machine learning · Digital history · Database · Printing privileges

Already in antiquity instruments played an important role in the study and practice of astronomy, as described in Chapter 2. The visualization and calculation of the movements of heavenly bodies were made possible with the help of more or less complex instruments. With the advent of printing in Europe, astronomical knowledge became accessible to a broader audience, as the process of printing enabled the production of much larger numbers of books than manuscripts, which were laboriously copied by hand. Printing also brought costs down, so that books became affordable to a more diverse readership that now had access to a brief and understandable description of basic astronomical knowledge with the help of books such as *De sphaera*. Though books were still not accessible or affordable for all, instruments made from sturdy materials such as brass or ivory were beyond the financial means of the most relevant targeted audience of the *Sphaera*: students.

4.1 The Function of Volvelles

Early modern printing techniques allowed a return of moveable parts in books as an approach that derived from antique sources (Karr 2004, 101). Moveable parts such as flaps and volvelles extended the possibilities of two-dimensional books and, as Lindberg put it, "challenged the bibliographical boundaries" (1979, 49). As shown in Chapter 2, volvelles served as mnemonic and pedagogical support or as a means of obtaining *new* knowledge, as in Ramón Llull's or Jacobus Publicius' cases.

One of the most popular examples of the astronomical volvelle are the ones from *Astronomicum Caesareum*, by the astronomer Petrus Apianus (1501–1552). Apianus created the elaborate and beautifully designed work for Emperor Charles V (1500–1558) and his brother, apparently adopting the astrolabe for the design of his volvelles. The book contains different volvelles for determining the locations of the planets, the Sun and the Moon, as well as for eclipse calculations. The volvelles could have also been used for casting horoscopes and are based on the Ptolemaic model of the Universe. They are equipped with usually five and sometimes six hand-colored layers of discs and make the book, as Owen Gingerich states, a "precision instrument" (Gingerich 1971; Gislén 2016).

The volvelles that appear in the *Sphaera* corpus from the sixteenth century seem to be less complex than the ones from Apianus's work. This will be investigated in more depth in Chapter 6, where some of the volvelles will be examined, but this initial impression can help with trying to understand the printers' reasons for including them in a book that was already a successful seller for almost 300 years without moveable parts. As the focus of *De Sphaera* was neither the included volvelles nor the making of these instruments, it is crucial to understand the background of their appearance in some of the *Sphaera* editions. At first glance, it is obvious that the volvelles were parts of the composition of the printed books and accompanied the text.

For an analysis of the volvelles' appearance, cultural context, and intended use, it is important to reiterate three points stated above:

1. The content of the *Sphaera* was shaped over centuries by the addition of different/new knowledge to it;
2. Volvelles had a kind of "golden age" in Europe around a century after the advent of printing;
3. Volvelles were used for, among other things, mnemonic purposes through visualization or the repetition of basic concepts.

In the following section the present state of all volvelles in the *De sphaera* corpus will be demonstrated through an analysis of the data from the research project The Sphere. The methods the data analysis used and which groups of volvelles were identified will be explained. Subsequently analysis will focus on a particular group of four volvelles from the *Sphaera* corpus, which I assign the name the "Wittenberg group." Extant volvelles in this group will be examined to create a comprehensive image of their condition, including:

- base discs and volvelle parts, which are analogous to "unassembled volvelles;"
- correctly assembled volvelles;
- incorrectly assembled volvelles.

An important aspect of the underlying data is the representativeness of the used corpus. The research project The Sphere used a single representative copy for each of the editions of the *Sphaera* included in the overall corpus. When considering the text or contents of books, there should not be any difference between the representative editions used and their copies. But when investigating the materiality of a copy and, particularly, the traces of its usage, which is the case with assembled or unassembled volvelles, the situation is somewhat different. Each copy was probably used by a different person and thus the usage of every copy and the respective volvelles was different. As a result, the underlying data used in this investigation can be only partly representative of the corpus and will not picture real and absolute numbers of volvelles in different states, but can give an estimation of the situation.

Nevertheless, this background will help in analyzing the material culture of the volvelles, which will be followed by a detailed description of all four volvelles.

In total, the *Sphaera* corpus contains twelve types of volvelle. Those volvelles that do not belong to the Wittenberg group will be described in the following section, but will not be analyzed in detail.

4.2 Situation Report: Volvelles

Joseph Klug (1490–1552), a printer from Wittenberg, began printing a series of *Sphaera* editions in 1531 that contained a recurring composition of images for five editions published and printed by him.[1] In 1538 he added an interesting feature to the *Sphaera* treatises: volvelles (Melanchthon and Sacrobosco 1538). Including Joseph Klug's 1538 edition, 123 editions containing various volvelles were printed in Europe, the last in 1647 by Bonaventura and Abraham Elsevier (fl. 1622–1652) in Leiden (Burgersdijk and Sacrobosco 1647), of a total of 359 editions. Thus, about 34% of the *Sphaera* editions between the years 1472 and 1656 contain volvelles (Fig. 4.1). According to Owen Gingerich's and Suzanne Karr Schmidt's calculations, about 123,000 copies of *Sphaera* editions with volvelles must have existed. The project's corpus of copies alone contains 593 volvelle pieces—which includes

---

[1] The Sphaera editions printed by Joseph Klug in Wittenberg can be found in the Sphaera database: http://hdl.handle.net/21.11103/sphaera.100802.

Fig. 4.1 *Sphaera* corpus—editions with and without volvelles. Plot by the author

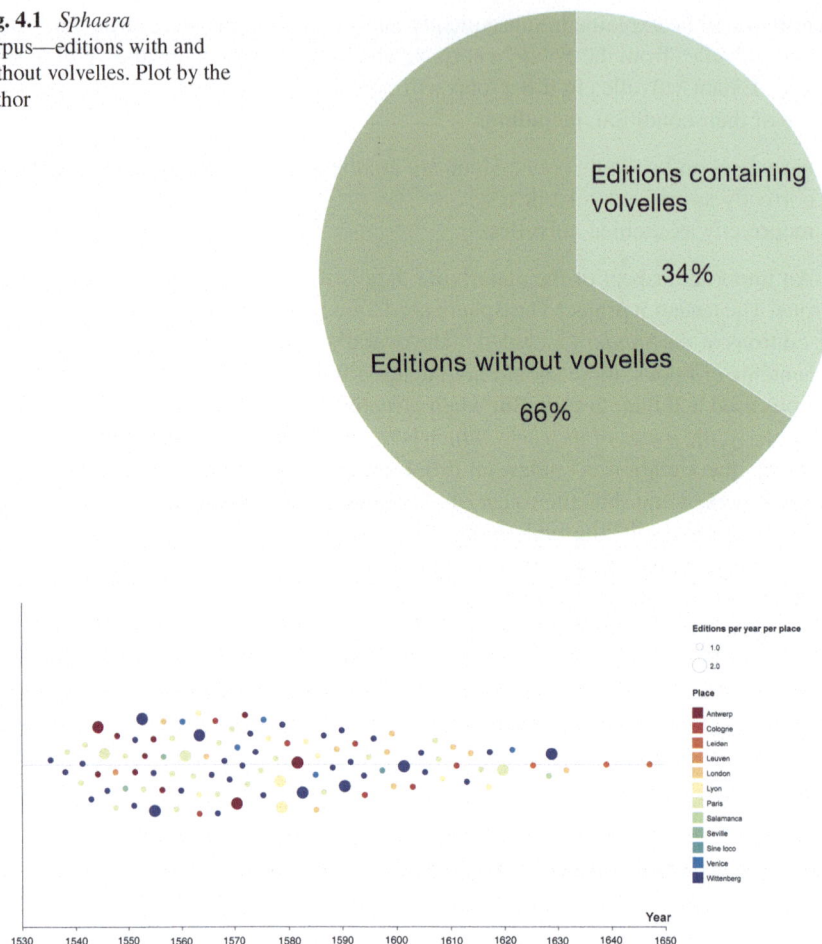

Fig. 4.2 Editions of the *Sphaera* containing volvelles ordered by printing date. Plot by the author[3]

assembled volvelles but also parts such as base discs or pointers (Fig. 4.2).[2] The following sections will describe how this data was generated and which groups of volvelle were identified with the help of this method.

---

[2] At this point it is quite important to repeat that, while each copy is representative of its edition, the whole corpus of copies that makes up this database is not representative of the early modern printing practice of the *Sphaera* itself. The copies were picked for practical reasons. However, the high number of volvelle pieces and volvelles as such can give an idea of the culture around volvelles in the *Sphaera* corpus.

[3] All plots were made with RawGraphs (https://rawgraphs.io) and Adobe Illustrator.

4.2 Situation Report: Volvelles

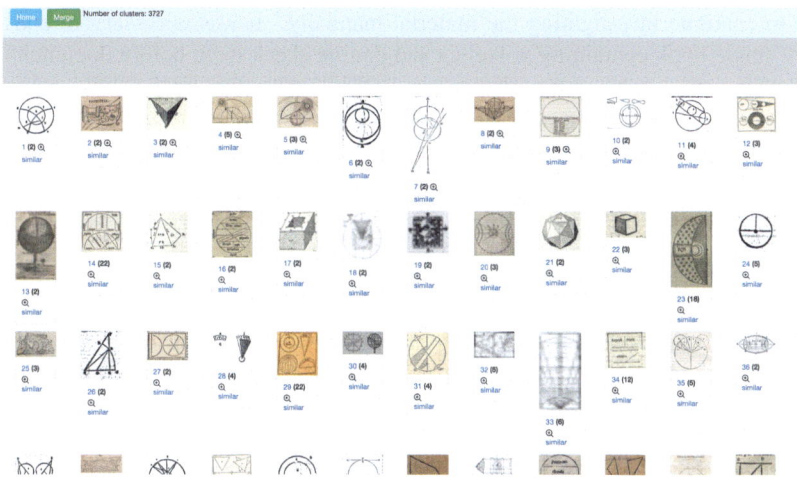

Fig. 4.3 Insight of the Sphaera Infrastructure Tool. Screenshot by the author

4.2.1 Data Collection

The collection, identification, and examination of this large amount of historical data, especially given that it contains not just numerical data but also images, can be handled best with the help of computational methods. As described previously, the 359 *Sphaera* editions and their data are managed in the research project The Sphere with the help of a database (see Sect. 3.2). It was stated that it was possible to capture all images in every edition for the database with the help of a specially created tool. This tool, called CorDeep, classifies visual elements with the help of a machine learning algorithm and is able to extract them from books (Kräutli et al. 2020; Büttner et al. 2022).[4]

To provide an environment tailored for working on the images, a separate app, the Sphaera Infrastructure Tool, was developed, and all the images were transferred to it. Within this app it is possible to compare the images and group them manually according to a desired principle, as, for example, resemblance (Fig. 4.3). With the Sphaera Infrastructure Tool it was possible to compare different stages of volvelles and how and if they were assembled. In this way it was also feasible to identify volvelles that were interpreted as images before the launch of the app, as their pointers and discs were shown in different settings.

During my research I grouped all volvelles and volvelle pieces within the tool in several runs to ensure that all were captured. The project's data scientist Hassan El-Hajj extracted the images and related data from my created clusters so that I was

---

[4] The tool CorDeep was launched in 2022 for the public and can be used to trace and identify visual elements in uploaded books: https://cordeep.mpiwg-berlin.mpg.de, accessed 28 August 2024.

able to continue investigating the material manually.[5] It was necessary to consult every single book containing volvelles and double check them before documenting the data in various tables for further research. While investigating the books, images, and metadata closely, it was possible to identify three volvelle groups, which will be described in the following section.[6]

4.2.2 Volvelle Groups

Within the corpus of *Sphaera* editions it was possible to identify three different groups of volvelles: The first group of volvelles appeared in 1538 in Joseph Klug's Wittenberg print and was subsequently printed for 91 years in various cities (Fig. 4.4). It consists of four volvelles that accompany the original structure and topics of the *Sphaera* (Fig. 4.5). As it was first and mainly printed in Wittenberg, I call it the "Wittenberg group." It makes up the largest grouping of the volvelles of the *Sphaera* corpus. This outcome tallies with the findings of Zamani et al. (2020, 2), who identified a family of treatises, a so called "epistemic community," that was produced in Wittenberg from the 1530s. Wittenberg at that time was an influential center of production, whose innovations were observed and adopted by other production centers internationally and thus disseminated within editions of books that were not produced in Wittenberg (Zamani et al. 2020, 11).

I was able to identify 109 editions containing the Wittenberg volvelles, which is about 30% of all *Sphaera* editions and 89% of all editions containing volvelles (Fig. 4.6). The Wittenberg group was printed for the last time in 1629 in three different editions: one edition in Salamanca by Jacinto Taberniel (fl. 1620–1642) and two in Wittenberg by Hiob Wilhelm Fincelius (fl. 1621–1666) and by the widow of Zacharias Schürer, Anna Schürer, with her daughters Anna Margareta and Maria Elisabeth (fl. 1626–1640) (Clavius et al. 1629; Blebel and Regiomontanus 1629; Melanchthon and Sacrobosco 1629).

Whether these four volvelles were integral to the text and whether they had a mnemonical or pedagogical function or a completely different purpose will be investigated in Chapter 6. In the following sections the focus is on the Wittenberg group, followed by an analysis of the material culture in Chapter 5, which is necessary to understand the emergence and diffusion of volvelles.

The second cluster of volvelles, the Seville group, appeared first in 1551 in Seville in a text by Martin Cortés under the title *Breve compendio de la sphera y de la arte de navegar* (Cortés 1551). It consists of six volvelles accompanying a *Sphaera* adaption[7]

---

[5] For more information on Hassan El-Hajj's work, see https://www.mpiwg-berlin.mpg.de/users/hhajj.

[6] The research data and corpus used for this work is published here: https://doi.org/10.5281/zenodo.15493638.

[7] For the category "adaption" and how it is defined in the context of the project *The Sphere*, see Sect. 3.2.

4.2 Situation Report: Volvelles

Fig. 4.4 Places of printing and numbers of editions of the first volvelle group. Plot by the author

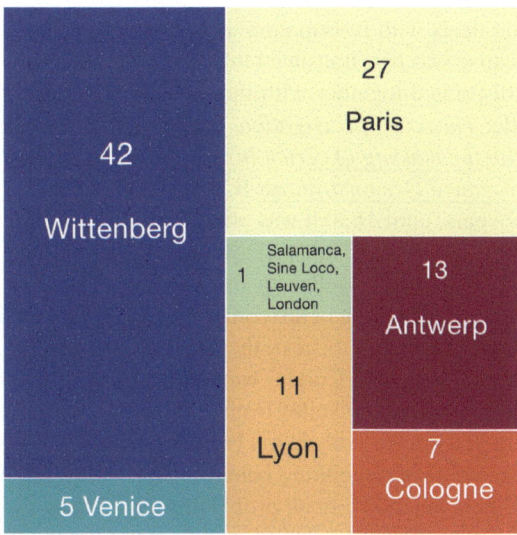

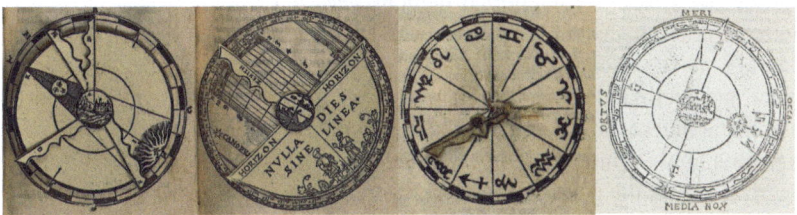

Fig. 4.5 The four volvelle types from the Wittenberg group, taken from different sixteenth-century editions. BSB München, Astr. U 158, urn:nbn:de:bvb:12-bsb10173685-3; Library of the Max Planck Institute for the History of Science

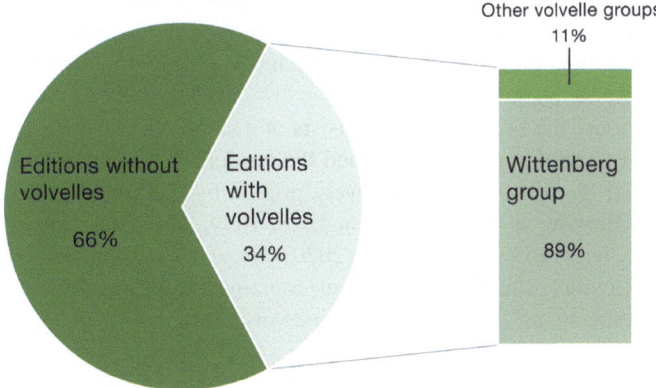

Fig. 4.6 The Wittenberg group inside the *Sphaera* corpus. Plot by the author

that deals with information about astronomy for navigational purposes. When this treatise was first translated into English by Richard Eden in 1561 the volvelles were still printed together with the text and were also referred to as instruments in the title: *The Arte of Navigation, Conteynyng a compendious description of the Sphere, with <u>the makyng of certen Instrumentes and Rules for Navigations: and exemplified by manye Demonstrations</u>* (Cortés 1561).[8] From 1551 this treatise was printed for 79 years, until 1630 it was printed for the last time in London by Bernard Alsop (fl. 1615–1652) and Thomas Fawcett (fl. 1625–1655) (Tapp and Cortés 1630).

All of the volvelles in these treatises, despite having different functions, can support navigation or are connected to it. The volvelles, referred to as "instruments," are introduced in the text, their function is described, and a workflow on how to use them is supplied. Cortés' work supports a hands-on learning experience; however, as many of the volvelles from the Cortés editions are assembled incorrectly, deconstructed or uninstalled, it is beyond the scope of this thesis to investigate the correct use and material culture behind them. Nevertheless, I will give a brief explanation of their functions. In all probability Fig. 4.7a–f shows the correct installation of the volvelles. The volvelle in Fig. 4.7a is a common form that we also find in the other two groups of *Sphaera* volvelles: It demonstrates not just the sun's daily motion through the year but also the relation between the height of the pole and the horizon. That in Fig. 4.7b is used to calculate the moon's position and days in connection to the place and declination of the sun. The volvelle in Fig. 4.7c is used to demonstrate an instrument that can tell the time when used properly, while that in Fig. 4.7d should demonstrate how to find northeast with the help of a compass. The volvelle shown in Fig. 4.7e shows the position of the North star and that in Fig. 4.7f the rising and setting of the sun (Cortés 1584).[9]

As navigational treatises, the Cortés editions were not meant to be used in university contexts and did not target students as their audience. My conclusion is that this could be a reason why the volvelles in this group were often not installed or assembled incorrectly: there was either no need to install the volvelles for pedagogical uses or more sturdy instruments existed that helped with the calculations that could have been made with the help of these volvelles which were than preferred to use over the paper version in the book.

The final group, the Leiden group, consists of a single volvelle that was printed just a few times and never made it beyond the borders of Leiden in the boundary of the corpus (Fig. 4.8). It was exclusively printed by the nephews Bonaventura and Abraham Elsevier in four editions in 1626, 1639, and 1647 (Burgersdijk and Sacrobosco 1626a, b, 1639, 1647; Buning 2020). These works were the only *Sphaera* editions the Elseviers printed and all of them contained volvelles. Content-wise, the volvelle, like others—as, for example, the first volvelle of the Seville group and the

---

[8] Author's emphasis.

[9] In four English editions of *The Arte of Navigation* volvelle 4e is replaced by 4d. As a seventh volvelle another volvelle is added to the work, of which I have not yet been able to find a complete version; therefore, this volvelle will not be described here.

4.2 Situation Report: Volvelles

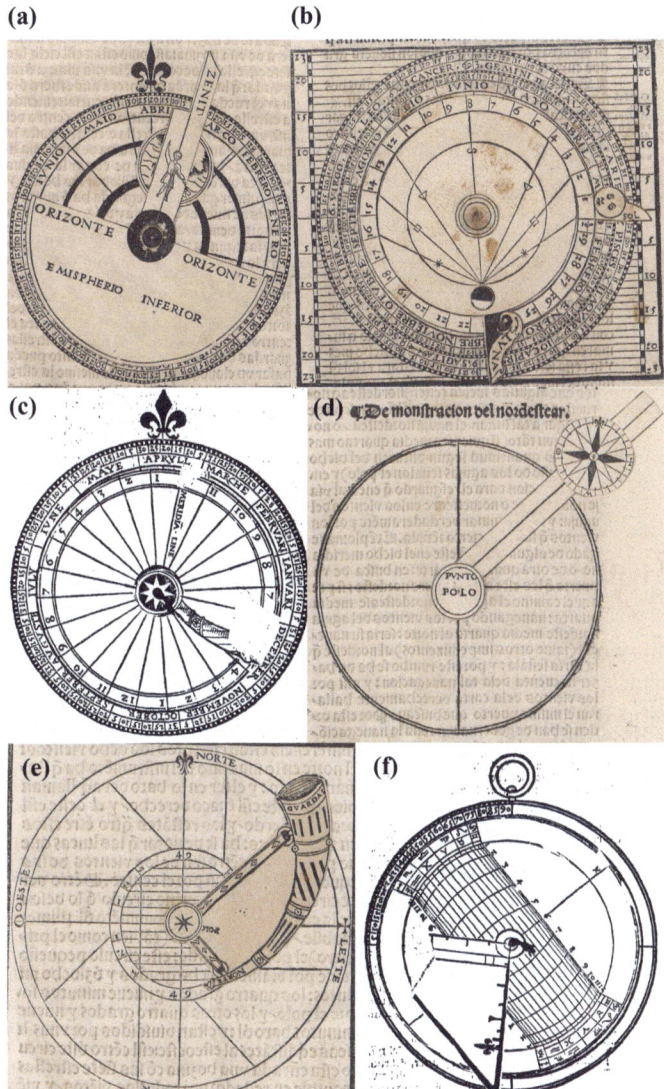

Fig. 4.7 a–f The six volvelles (a–f from the upper left, line by line) from the Seville group. From Cortés (1551), Biblioteca Nacional de España, R/1683 (a, b, d, e) and Cortés (1579), New York Public Library (c, f)

second volvelle from the Wittenberg group—deals with the position of the horizon, demonstrating the relationship between the horizon and the zodiac signs.

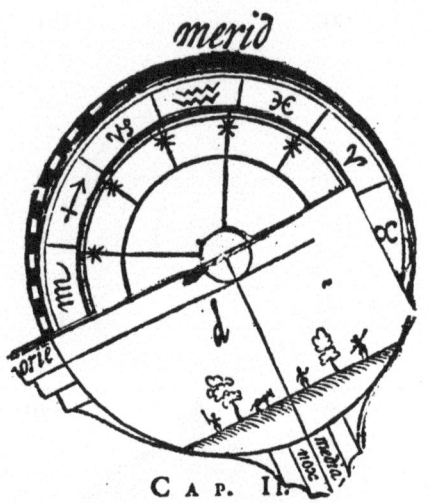

Fig. 4.8 The Leiden volvelle. From Burgersdijk and Sacrobosco (1647). Université de Lausanne, Public Domain Mark

4.2.3 Privilegia impressoria *in the Context of the* Sphaera

Although this work focuses on the study of the knowledge embedded in the volvelles of the Sphaera, it is important to briefly consider an aspect of the early modern printing industry that is often relegated to the background, neglected in research, or treated in separate studies: *Privilegia impressoria*. The codification of practical knowledge that shaped Europe's scientific identity did not take place in a vacuum, isolated from its environment, but was subject to legal rules, control, access, and authority. The dissemination of this knowledge required a complex distribution system in which competition played a significant role, especially with regard to a reliable outlet such as university textbooks. One way of circumventing such competition and repositioning oneself in the market was to reprint books and sell them at a lower price than the competition. The history of the book trade is replete with stories of reprinting and fighting reprinting. Printing a book in the early modern period was a complex process that required advanced skills and resources on many levels. Not only time, but also money and labor were invested in book production. To avoid these costs and to save time, it was a welcome situation for a printer to simply take a book already printed by another printer and produce an exact copy of it, possibly on cheaper paper. In doing so, the printer avoided several steps in the process, such as proofreading, calculating the number of pages, placing the images in the right place, etc. The original printer, publisher, or author suffered a loss of income and economic damage. Another advantage of reprinting was that the needs of the market could be better assessed. To avoid this kind of book copying and to protect their work, printers, authors, and publishers could apply to the authorities for a so-called *Privilegium Impressorium*, a printing or book privilege that protected their works from such piracy. Privileges were not only granted to printers—booksellers, publishers, authors or engravers could also protect their works in this way.

Book privileges are considered an early form of today's copyright.[10] With the first privilege granted in 1479, more and more countries granted privileges to their printers, publishers, authors and engravers. In what is now Germany, privileges appeared occasionally in the sixteenth century (Armstrong 2002). This trend evolved and culminated in a complex system with several authorities involved, such as the imperial book commissioner in Frankfurt am Main, who was responsible for monitoring the regulations. Such privileges provided effective protection. For example, a privilege granted by the Holy Roman Emperor to the German Nation had a scope that covered the entire empire and prohibited the reprinting of the book or works listed in the privilege (Koppitz 2008, VIII). The system of authorities who could grant such privileges was and is a complex matter involving several levels of jurisdiction. Not only the highest authorities, such as the Emperor or the Crown, could grant such privileges, but also lesser jurisdictions, such as the Electorate of Brandenburg or the Privy Council of the Habsburg Netherlands, had the legal authority to interact with the printing industry in this way. In addition, local printers were careful to regulate reprints or transactions with authors to prevent reprints or disputes over claims at the local level. The *Frankfurter Buchdruckerordnung* (printer's ordinance), for example, contained several pages detailing how reprints were regulated at this local level.[11] Because of the influence of the privilege system on the early modern knowledge economy, the factor of "privileges" should be mentioned at least briefly when discussing the spread of written knowledge and volvelles in this period.

In order to find out whether book producers requested privileges for their printed *Sphaera* works, the books in the corpus used for this work were searched for such traces with the help of the database. In most cases, printers were advised to print at least an excerpt of the original privilege text in the book, which can still be found today. Depending on the region and situation, privileges were often valid for three, six, or ten years with exceptions-some were even valid for the lifetime of the printer. Since privileges were also considered good advertising, printers would continue to print "Cum privilegio" on the cover of the book even after the privilege had expired. In the case of the underlying corpus of volvelle books, printers, publishers or authors who held privileges for these books were mainly located in Paris (31%), Antwerp (24%) or Lyon (24%) (Fig. 4.9). This is not surprising, since not only were most of the *Sphaera* works printed in Paris, but also a high proportion of the volvelle editions were printed in Paris (see Sect. 4.2.2). In general, a quarter of the Volvelle editions were privileged by authorities, in over 50% of the cases by the French crown.

However, what is particularly noteworthy about printing privileges in the context of the present corpus is the absence of things, or what we do not see. In the list

---

[10] The matter of this early form of copyright is investigated in the framework of the project Before Copyright (ERC, BE4COPY, 101042034) at the University of Oslo, Norway. "The researchers in the BE4COPY project study the changing nature of the printing privilege over the course of [...] 300 years. The intimate relationship between legal frameworks and the politics of knowledge is the primary focus of the project." https://www.hf.uio.no/iakh/english/research/projects/before-copyright/about/.

[11] ISG H.10.02 Nr. 30. 1598. *Eines Erbarn Raths ernewerte Ordnung und Artickel / wie es forthin auff allen Truckereyen / in dieser Statt Franckfurt soll gehalten werden,* fol. Aiii.

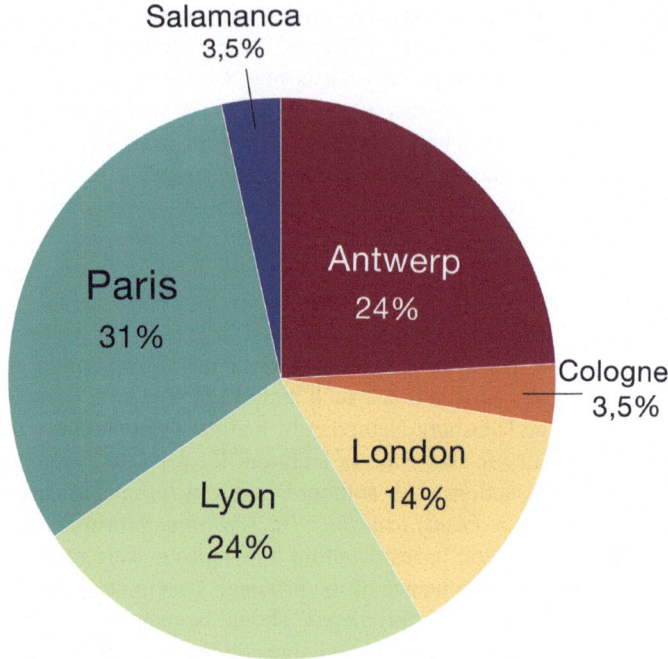

Fig. 4.9 Printing places of books containing volvelles and printing privileges. Plot by the author

of privileges for books with volvelles, only one privilege was granted to a person in Cologne. This is the only one granted in the jurisdiction of the Holy Roman Empire of the German Nation. Notably, Wittenberg, where the Wittenberg group of volvelles originated, is missing. This is interesting when one considers the "epistemic community" with which the Wittenberg group is undeniably associated. As noted above, the editions printed in Wittenberg fostered this epistemic community, and Wittenberg was an influential production center whose innovations were observed and adopted by other production centers internationally. As Zamani et al. (2020, 12) noted, Wittenberg, Antwerp, and Paris were linked as the three main centers of production and distribution of this specific epistemic community, which also produced the Wittenberg group of volvelles. Yet not a single book of volvelles printed in Wittenberg contained a privilege. Since the Wittenberg composition was first printed in Wittenberg in 1538 and was printed 24 times with privileges under the French Crown between 1547 and 1583, it can be assumed that the privileges of the French Crown are an additional factor that should be taken into account when discussing the success of the Wittenberg Composition. Ultimately, these privileges contributed to an acceleration in the distribution and transmission of books containing this Volvelle composition.

4.3 The Wittenberg Group

As stated in this chapter, the Wittenberg group was the first cluster of volvelles integrated within printed editions of the *Sphaera*, as well as being the group of volvelles that was most often copied within the corpus, and most often privileged by authorities. First appearing in 1538 in Wittenberg, these volvelles were reprinted—initially via Antwerp and Paris in 1543—109 times until 1629 in various Western European cities. Most of the editions, however, were from Wittenberg (Fig. 4.10).

To build the database for this group all volvelle pieces and assembled volvelles were clustered and analyzed. I was able to identify 512 individual pieces that belong to the Wittenberg group, making this group a particularly interesting foundation for research. Each volvelle consists of the following pieces:

- A "base disc," which is printed on a page inside the book. It shows where the volvelle is meant to be assembled and sometimes holds information that is important for the volvelle and supplements its function;
- "Parts," which are meant to be attached to the base discs to form a complete volvelle. Parts can be pointers or discs. Sometimes a volvelle consists of several discs and pointers, sometimes it just holds a single pointer attached to a base disc.

The largest subdivision of the 512 pieces in this group is the base discs, of which 285 exist. The second largest subdivision is the above-described "parts," followed by ninety-six correctly assembled volvelles and twenty-four incorrectly assembled volvelles (Fig. 4.11). All of these subdivisions will be described in the following sections. Although correctly assembled volvelles would seem especially interesting

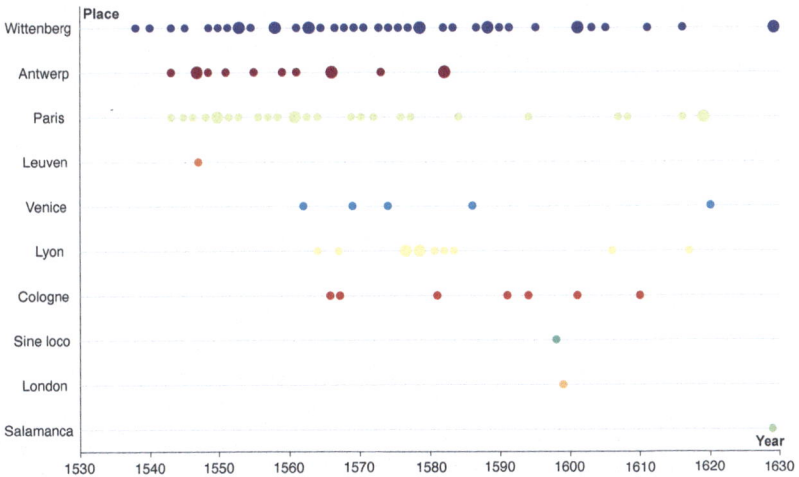

Fig. 4.10 Printing history of the Wittenberg group. Plot by the author

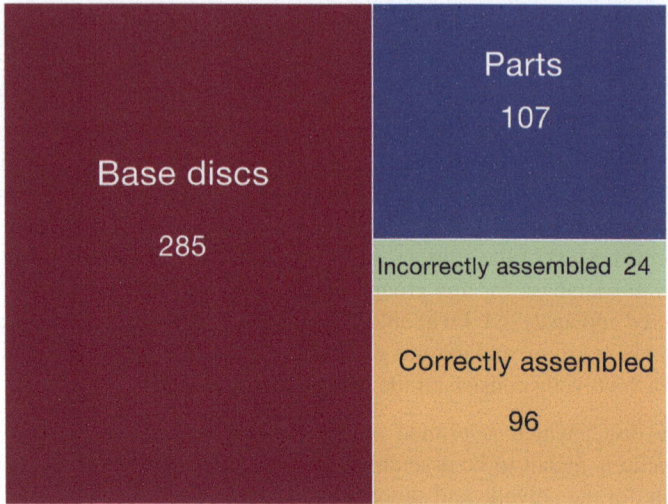

Fig. 4.11 Conditions of Wittenberg group volvelles and their parts. Plot by the author

for investigating the functions of these instruments in the *Sphaera* corpus, incorrectly assembled pieces, parts, and base discs, which are analogous to unassembled volvelles, are also very informative about the volvelles' history and material aspects.

4.3.1 Base Discs and Parts

Base discs and volvelle parts, such as pointers and discs, account for most of the volvelle corpus from the *Sphaera* editions. As previously described, it was the reader's responsibility to install volvelles if desired or necessary, and it seems probable that students in universities assembled the volvelles together with professors.

In reality, most of the printed volvelles were never installed, a conclusion derived from the number of remaining base discs. This could be because unsold copies were the ones that were kept and conserved until today. In these cases, no reader was in charge to assemble volvelles, while many of the copies that were sold and thus used may have been lost as a result of their usage. In addition to the volvelles that were never installed, some installations were reversed and some were performed incorrectly. These volvelles also just show a base disc and are thus counted within the subdivision of base discs.

The base discs and thus unassembled volvelles can tell us three main things:

- how many editions contain volvelles;
- how the base disc looks, when it is not hidden by other volvelle parts;
- where the volvelle should have been installed.

4.3 The Wittenberg Group

The last aspect is especially interesting because each of the *Sphaera*'s four chapters deals with distinct topics that can give references on the use of the volvelles by the subject matter of the chapters. This matter will be treated in Chapter 6, where the functions of the four volvelles of the Wittenberg group are described.

As described in Chapter 2, the volvelle parts that were sold together with a book were usually printed on a double sheet and bound in the back of the book. In many cases, the sheet containing the parts is lost. To understand how these sheets could have been lost and why they were not in some cases, I examined several editions in the library of the MPIWG and the Österreichische Nationalbibliothek (ONB; Austrian National Library), Vienna, Austria in order to understand both the binding technique and the installment methods of the volvelle discs. Although the books were bound after being sold, it is possible that the part-sheet wasn't bound together with the book so the parts could more easily be cut out. However, two editions were identified where the part-sheet was printed and bound in a way that hindered the cutting process: In one, the parts were printed on the back of a page containing text that belonged to the book (Vinet et al. 1569, 164–165), so that if the reader wanted to cut out and install the parts of the volvelles he or she had to decide whether the text was needed or not. In the second case the part-sheet was bound as the other pages of the book, cutting some pointers and discs in half (Hero et al. 1591, M3) and ensuring that the parts couldn't be taken out of the book without damaging them in a way that made it impossible to add them to a volvelle.

Beside these cases I found three remaining sheets in three *Sphaera* editions from Lyon and Wittenberg that contain step-by-step instructions on how to construct the volvelles correctly. On these sheets all the parts of the book's volvelles—described as "wheels"—were printed together and marked with letters to assist in their assembly. Figure 4.12 shows the sheet containing pieces of the four volvelles of the Wittenberg group, to which, according to their appearance and intended use, I have assigned the following terms: eclipse volvelle, horizon volvelle, zodiac volvelle, and heliacal volvelle (Giuntini et al. 1581, 654–655; Peucer 1558; Melanchthon and Sacrobosco 1629, A14). All three of the books concerned were printed in the octavo format, which means a folio sheet was printed and then folded three times to create eight leaves (sixteen pages) inside a book. To fit the volvelle parts and the textual description on one page inside the book, a whole folio sheet was used and bound as a regular page. After the book was bound, the folio sheet with the volvelle parts was folded three times to make it as small as a normal octavo page.

Below, parts of the descriptions are translated:

From these types make wheels on the two figures in the first chapter of the Sphaera

Their notes are E, F [...]

At the beginning cut out the other types [*parts*] from the paper, so that they become more suitable to use [...]

The moon will be applied to the figure of the first chapter E, so that the horizon F, H, G may also be applied to the earth E, E [...]

In the same manner Horizon F, [...] is adjusted to the figure F of the first chapter [...]

The wheels with the note O are adapted to the figure of the ortu Poetico in the beginning of the third chapter, so that the earth O remains unmoved [...]

50   4 Volvelles in the *Sphaera* Corpus

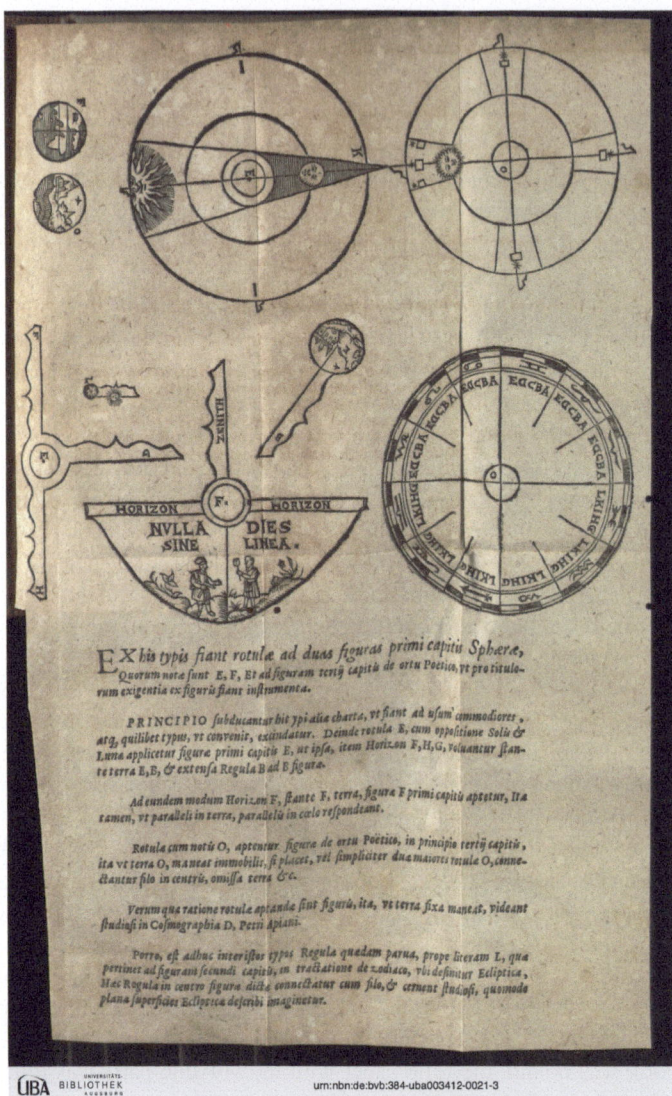

Fig. 4.12 Folio sheet with volvelle parts and descriptions. Parts marked "E" belong to the eclipse volvelle, parts marked "F" to the horizon volvelle, the pointer marked "L" belongs to the zodiac volvelle, and parts marked "O" belong to the heliacal volvelle. Annotations by the author. From Melanchthon and Sacrobosco (1629, A14). Public Domain Mark, Universitätsbibliothek Augsburg

4.3 The Wittenberg Group 51

> Besides there is a small rule among these types, near the letter L, which belongs to the figure of the second chapter in the treatment of the zodiac [...]
>
> The rule is connected with a thread to the center of the figure and the students will see how to imagine and describe the planes of the ecliptic surface.[12]

All the descriptions from the three editions described above contain identical text, although the edition printed in Lyon refers not just to the chapters where the base discs can be found but also to the page numbers (Giuntini et al. 1581). The instructions clearly name students as the intended users of the volvelles. With the help of these instructions, it is possible to identify which of the complete volvelles extant within the corpus are correctly and incorrectly assembled. The former will be described in the following section.

4.3.2 Correctly Assembled Volvelles

The database provided by the project gives an overview of the assembled volvelles within the corpus. It was stated above that the corpus examined here is contains only one representative copy of each edition; therefore, there are many more copies in existence that could be analyzed if available and, though the database is not reliable for all early modern *Sphaera* copies, they can illustrate trends in volvelles' assembly and use.

As stated in Sect. 4.3, about ninety-six of the 512 volvelle pieces from the Wittenberg group belong to the category "Correctly Assembled Volvelles." Unlike those of the Seville group, the Wittenberg volvelles are not described in the editions' text and their construction is also not described in most cases, except for the three pages containing instructions that were described in Sect. 4.3.1.

The distribution of correctly assembled volvelles found within the corpus can give insights on how often each type of volvelle was assembled correctly and also which type of volvelle was least often assembled correctly (Fig. 4.13). The distribution of base discs and thus of unassembled volvelles, is nearly the same for all four volvelle

---

[12] Melanchthon and Sacrobosco (1629, A14): *EX his typis fiant rotulae ad duas figuras primi capitis Sphaerae, Quorum notae sunt E, F, Et ad figuram tertij capitis de ortu Poetico, vt pro titulorum exigentia ex figuris fiant instrumenta. PRINCIPIO subducantur hitypi alia charta, vt fiant ad usum commodiores, atque quilibet typus, vt convenit, excindatur. Deinde rotula E, cum oppositione Solis & Lunae applicetur figurae primi capitis E, ut ipsa, item Horizon F, H, G, voluantur stante terra E, E, & extensa Regula B ad B figurae. Ad eundem modum Horizon F, stante f, terra, figurae F primi capitis aptetur, Ita tamen, vt paralleli in terra, parallelis in coelo respondeant. Rotulae cum notis O, aptentur figurae de ortu Poetico, in principio tertij capitis, ita vt terra O, maneat immobilis, si placet, vel fimpliciter dua maiores roltuae O, conectantur filo in centris, omissa terra & c. Verum qua ratione rotulae aptandae sint figuris, ita, vt terra fixa maneat, videant studiosi in Cosmographia D, Petri Apiani. Porro, est adhuc inter istos typos Regula quaedam parua, prope literam L, quae pertinent ad figuram secondi capitis, in tractatione de zodiac, vbi definitur Ecliptica, Haec Regula in centro figurae dictae connectatur cum filo, & cernent studiosi, quomodo plana superficies Eclipticae describe imaginetur.*

types, although about 10% more base discs of zodiac volvelles remain than of the other volvelle types.

The zodiac volvelle, because of its simplicity, was probably either often misinterpreted as a single image or not assembled because the reader didn't feel that its assembly was necessary. On the other hand, this volvelle is—together with the horizon volvelle—most often assembled correctly (Figs. 4.12 and 4.14), perhaps because it was simple: it consists of only a base disc and a single pointer (Fig. 4.12). The horizon volvelle, as the second simplest volvelle—consisting of just one additional disc besides the base disc and an optional centerpiece—also wouldn't result in many errors during the construction process.

As Fig. 4.13 shows, the so called heliacal volvelle was least often assembled correctly. This is apparently due to the fact that it was—and still is—not clear which

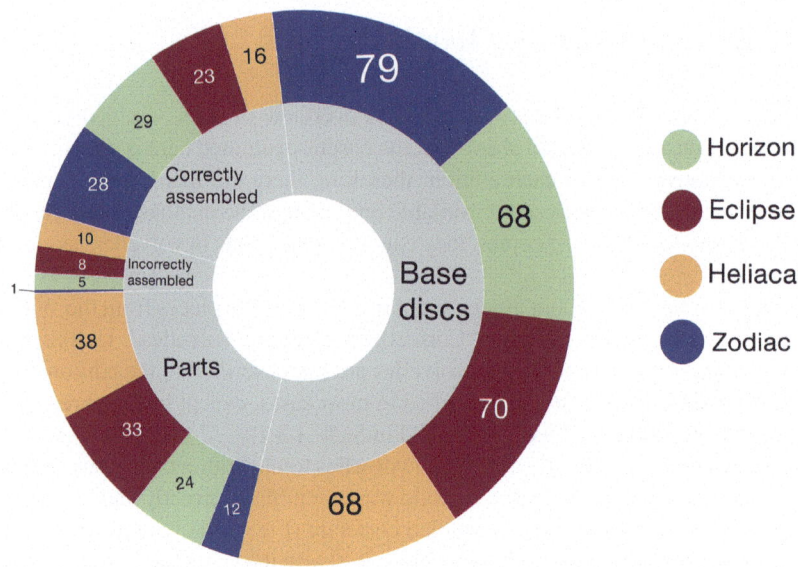

Fig. 4.13 Distribution of volvelle types according to condition. Plot by the author

Fig. 4.14 a–d Left to right: Eclipse volvelle, horizon volvelle, zodiac volvelle, heliacal volvelle. Images taken from different *Sphaera* copies from the sixteenth century. BSB München, Astr. U 158, urn:nbn:de:bvb:12-bsb10173685-3; Library of the Max Planck Institute for the History of Science

parts of the third disc (showing the moon and cut-out squares) should be cut out, even when instructions are given (Fig. 4.12). In addition, the complexity of the topic (described in Sect. 6.4) of this volvelle may have caused irritation for the student reader of this part of the work.

As the distribution of correctly and incorrectly assembled volvelles shows, the least "complex" volvelles (consisting of fewer than three parts) were most often assembled correctly, while the most "complex" volvelles were more often assembled incorrectly. This does suggest a correlation between the complexity of the instrument and the likelihood that it would be assembled correctly—or at all.

4.3.3 Incorrectly Assembled Volvelles

It is likely that most of the part-sheets provided lacked the previously described step-by-step instructions, and that the construction of the volvelles on those cases would have been a trial-and-error process, a process that needed previous knowledge of these volvelles, or a collective process in which students and professor collaborated. The heliacal volvelle, consisting of four parts with an additional cut-out-process, was most frequently constructed incorrectly, as intimated above.

After this overview of the various groups of *Sphaera* volvelles, as well as further insight into the Wittenberg group, including its temporal and spatial distribution, Chapter 5 will now examine the material culture context, followed in Chapter 6 by the descriptions of the individual volvelles.

Bibliography

Digital Repositories

Sphaera Corpus*Tracer*   Max Planck Institute for the History of Science. https://db.sphaera.mpiwg-berlin.mpg.de/resource/Start.

Archives

ISG   Institut für Stadtgeschichte, Frankfurt am Main, Germany
MPIWG   Library of the Max Planck Institute for the History Science, Berlin, Germany
ONB   Österreichische Nationalbibliothek (Austrian National Library), Vienna, Austria

Primary Sources

Blebel, Thomas, and Johannes Regiomontanus. 1629. *De sphaera, seu primi mobilis rudimentis libellus. Ad usum scholarum maximè accomodatus: accurata methodo & brevitate conscriptus à M. Thoma Blebelio Budissino. Et nunc ab infinitis propè mendis liberatus, tabulisque correctis instructus. Adiectus ad calcem est canon sinum Joh. Regiomontani, ad semidiametrum IOOOOOOO*. Wittenberg: Hiob Wilhelm Fincelius.

Burgersdijk, Franco, and Johannes de Sacrobosco. 1626a. *Sphaera Iohannis de Sacro-Bosco, decreto illustr. & potent. d d. ordinum hollandiae & west-frisiae, in usum scholarum ejusdem provinciae, sic recensita, ut & latinitas, & methodus emendata sit, multaque addita, qua ad huius doctrina illustrationem requirebantur. Operâ & studio Franconis Burgersdicii*. Leiden: Bonaventura & Abraham Elsevier. http://hdl.handle.net/21.11103/sphaera.101041.

Burgersdijk, Franco, and Johannes de Sacrobosco. 1626b. *Sphaera Iohannis de Sacro-Bosco, emendatiore sermone & methodo tradita, multisque praeceptionibus quae ad illustrationem hujus doctrinae requirebatur, adaucta. Operâ & studio Franconis Burgersdicii*. Leiden: Bonaventura & Abraham Elsevier.

Burgersdijk, Franco, and Johannes de Sacrobosco. 1639. *Sphaera Iohannis de Sacro-Bosco, decreto illustr. & potent. d d. ordinum hollandiae & west-frisiae, in usum scholarum ejusdem provinciae, sic recensita, vt & latinitas, & methodus emendata sit, multaque addita, quae ad huius doctrinae illustrationem requirebantur. Operâ & studio Franconis Burgersdicii*. Leiden: Bonaventura & Abraham Elsevier.

Burgersdijk, Franco, and Johannes de Sacrobosco. 1647. *Sphaera Johannis de Sacro-Bosco, decreto illustr. & potent. d d. ordinum hollandiae & west-frisiae, in usum scholarum ejusdem provinciae, sic recensita, vt & latinitas, & methodus emendata sit, multaque addita, quae ad hujus doctrinae illustrationem requirebatur. Operâ & studio Franconis Burgersdicii*. Leiden: Bonaventura & Abraham Elsevier. http://hdl.handle.net/21.11103/sphaera.100590.

Clavius, Christoph, Francesco Giuntini, Élie Vinet, and Johannes de Sacrobosco. 1629. *Exposicion de la esfera de Iuan de Sacrobosco doctor parisiense. Traduzida de Latin en lengua vulgar, augmentada y enriquecida, con lo que della dixeron Francisco Iuntino, Elias Veneto, Christoforo Clavio, y otrossus expositores, y comentadores. Por F. Luys de miranda de la orden de san francisco, lector jubilado, y provincial que ha sido, de la provincia de santiago, consultor del supremo consejode la santa general inquisicion. Dirigida al serenissimo señor cardenal infante D Fernando Arcobispo de Toledo, y primado de las Españas*. Translated by Luis de Miranda. Salamanca: Jacinto Taberniel.

Cortés, Martín. 1551. *Breve compendio de la sphera y de la arte de navegar, con nuevos instrumentos y reglas, exemplificado con muy subtiles demonstraciones: compuesto por Martin Cortes natural de burjalaroz en el reyno de Aragon y de presente vezino de la ciudad de Cadiz: dirigido al invictissimo monarcha Carlo quinto rey de las Hespañas etc. Señor Nuestro*. Seville: António Alvares.

Cortés, Martín. 1561. *The arte of navigation, Conteynyng a compendious description of the sphere, with the makyng of certen instrumentes and rules for navigations: and exemplified by manye demonstrations. Wrytten in the Spanyshe tongue by Martin Curtes, and directed to the emperour Charles the fyfte. Translated out of Spanyshe into Englyshe by Richard Eden*. London: Richard Jugge.

Cortés, Martin. 1579. *The arte of navigation, Conteynyng a compendious description of the Sphere, with the making of certayne instrumentes and rules for navigations, and exemplified by many demonstrations. Written by Martin Curtes, Spaniarde. Englished out of Spanyshe by Richarde Eden, and now newly corrected and amended in divers places. Whereunto may be added at the wyl of the byer, another very fruitefull and necessary booke of Navigation, translated out of Latine by the sayde Eden*. London: Widow of Richard Jugge.

Cortés, Martín. 1584. *The arte of navigation, Conteining a compendious description of the spere, with the making of certayne instrumentes and rules for navigations, and exemplified by many demonstrations. Written by Martin Cortes, Spaniarde. Englished out of Spanishe by Richarde*

Eden, and now newly corrected and amended in divers places. Whereunto may be added at the wyll of the byer, another very fruitefull and necessary booke of navigation, translated out of Latine by the sayd Eden. Translated by Richard Eden. London: Widow of Richard Jugge (Joan Jugge). http://hdl.handle.net/21.11103/sphaera.100675.

Giuntini, Francesco, Georg von Peuerbach, and Johannes de Sacrobosco. 1581. *Speculum astrologiae, comprehendens commentaria in theoricas planetarum, et in sphaeram Ioannis de Sacro Bosco: uná cum tabulis de eclipsibus Georgii Purbachii, et supputationibus motuum planetarum, secundum decreta Alphonsii regis Hispaniae: et Nicolai Copernici, cum diversis aliis tractatibus astrologicis. Autore Francisco Iunctino Florentino s.t.d. ac elecmosynario ordinario serenissimi principis Francisci Valesii, Henrici filij, Francisci nepotis, ac Christianiss Francorum, ac Poloniae regis fratris unici, andegauensis ducis, etc. Salvo per omnia iudicio sanctae sedis Apostolicae. Tomus posterior*. Lyon: Philippe Tinghi. http://hdl.handle.net/21.11103/sphaera.101207.

Hero, Albertus, Élie Vinet, Pierio Valeriano, Pedro Nunes, and Johannes de Sacrobosco. 1591. *Sphaera Ioannis de Sacrobosco emendata. Eliae Vineti Santonis scholia in eandem Sphaeram, ab ipso authore restituta. Quibus nunc accessere scholia Heronis. Adiunximus huic libro compendium in Sphaeram per Pierium Valerianum Bellunensem, Et Petri Nonii Salaciensis demonstrationem eorum, quae in extremo capite de climatibus Sacroboscius scribit de inaequali climatum latitudine, eodem Vineto interprete*. Cologne: Godwin Cholinus. http://hdl.handle.net/21.11103/sphaera.100326.

Melanchthon, Philipp, and Johannes de Sacrobosco. 1538. *Ioannis de Sacro Busto libellus, De sphæra: Eiusdem autoris libellus, cuius titulus est Computus, eruditissimam anni & mensium descriptionem continens. Cum praefatione Philippi Melanth. & novis quibusdam typis, qui ortus indicant*. Wittenberg: Joseph Klug.

Melanchthon, Philipp, and Johannes de Sacrobosco. 1629. *Libellus de Sphaera Johannis de Sacro Busto. Accessit ejusdem autoris computus ecclesiasticus, Et alia quaedam, in studiosorum gratiam edita. Cum praefatione Philippi Melanchthonis*. Wittenberg: Widow & heirs of Zacharias I. Schürer. http://hdl.handle.net/21.11103/sphaera.100303.

Peucer, Kaspar. 1558. *Elementa doctrinae de circulis coelestibus et primu motu*. Vitebergae: Crato. http://hdl.handle.net/21.11103/sphaera.100191.

Tapp, John, and Martín Cortés. 1630. *The art of navigation. First, written in the Spanish tongue by that excellent marriner and mathematician of these times, Martine Curtis. From thence translated into English by Richard Eden: And now newly corrected and inlarged with many necessary tables, rules, and instructions, for the more easie attaining to the knowledge of navigation. By Iohn Tap*. Translated by Richard Eden. London: John Tapp, Bernard Alsop, Thomas Fawcett.

Vinet, Élie, Pierio Valeriano, Pedro Nunes, and Johannes de Sacrobosco. 1569. *Sphaera Ioannis de Sacro Bosco, emendata. Eliae Vineti Santonis scholia in eandem Sphaeram, ab ipso authore restituta. Adiunximus huic libro compendium in sphaeram, per Pierium Valerianum Bellunensem, & Petri Nonij Salaciensis demonstrationem eorum: Quae in extremo capite de climatibus Sacroboscius scribit de inaequali Climatum latitudine: eodem Vineto interprete. Ex postrema Impressione Lutetiae*. Venice: Girolamo Scoto. http://hdl.handle.net/21.11103/sphaera.101040.

Secondary Sources

Armstrong, Elizabeth. 2002. *Before copyright: The French book-privilege system, 1498–1526*. Cambridge; New York; Port Chester: Cambridge University Press.

Buning, Marius. 2020. Fashioning cosmology: Franco Burgersdijk as the author of the dutch Tractatus de sphaera. In *De sphaera of Johannes de Sacrobosco in the early modern period: The authors of the commentaries*, ed. Matteo Valleriani, 359–389. Cham: Springer.

Büttner, Jochen, Julius Martinetz, Hassan El-Hajj, and Matteo Valleriani. 2022. CorDeep and the Sacrobosco dataset: Detection of visual elements in historical documents. *Journal of Imaging* 8 (10): 285.

Gingerich, Owen. 1971. Apianus's *Astronomicum Caesareum* and its Leipzig facsimile. *Journal for the History of Astronomy* ii: 168–177.

Gislén, Lars. 2016. A lunar eclipse volvelle in Petrus Apianus' Astronomicum Caesareum. *Journal of Astronomical History and Heritage* 19 (3): 247–254.

Karr, Suzanne. 2004. Constructions both sacred and profane: Serpents, angels, and pointing fingers in Renaissance books with moving parts. *The Yale University Library Gazette* 78 (3/4): 101–127.

Koppitz, Hans-Joachim. 2008. *Die kaiserlichen Druckprivilegien im Haus-, Hof- und Staatsarchiv Wien Verzeichnis der Akten vom Anfang des 16. Jahrhunderts bis zum Ende des Deutschen Reichs (1806)*. Ed. Hans-Joachim Koppitz, *Buchwissenschaftliche Beiträge aus dem Deutschen Bucharchiv München*. Wiesbaden: Harrassowitz Verlag.

Kräutli, Florian, Daan Lockhorst, and Matteo Valleriani. 2020. Calculating sameness: Identifying early-modern image reuse outside the black box. *Digital Scholarship in the Humanities* 36 (Supplement_2): ii165–ii174. https://doi.org/10.1093/llc/fqaa054.

Lindberg, Sten G. 1979. Mobiles in books. Volvelles, inserts, pyramids, divinations, and children's games. *The Private Library* 2 (2): 41–82.

Zamani, Maryam, Alejandro Tejedor, Malte Vogl, Florian Kräutli, Matteo Valleriani, and Holger Kantz. 2020. Evolution and transformation of early modern cosmological knowledge: A network study. *Scientific Reports* 10 (1): 1–15. https://doi.org/10.1038/s41598-020-76916-3.

Open Access This chapter is licensed under the terms of the Creative Commons Attribution 4.0 International License (http://creativecommons.org/licenses/by/4.0/), which permits use, sharing, adaptation, distribution and reproduction in any medium or format, as long as you give appropriate credit to the original author(s) and the source, provide a link to the Creative Commons license and indicate if changes were made.

The images or other third party material in this chapter are included in the chapter's Creative Commons license, unless indicated otherwise in a credit line to the material. If material is not included in the chapter's Creative Commons license and your intended use is not permitted by statutory regulation or exceeds the permitted use, you will need to obtain permission directly from the copyright holder.

Chapter 5
The Material Culture of the Wittenberg Group

Abstract To investigate historical artifacts it can be helpful to use methods developed in museal contexts. This chapter introduces the methods and concepts that will be applied in Chapter 6 to the volvelles from the corpus of editions of Johannes de Sacrobosco's *Sphaera*. As the volvelles are not images but rather instruments and can be manipulated, they were explored with the help of Edward McClung Fleming's so called Winterthur model, in which an artifact's five basic properties—history, material, construction, design, and function—are explored by following a workflow that consists of the operations identification, evaluation, cultural analysis, and interpretation. This methodology was used alongside Davis Baird's concept of "Thing Knowledge."

Keywords *Sphaera* · Johannes de Sacrobosco · Early modern period · History of science · History of knowledge · Volvelles · Material culture · Winterthur model

Things that exist in everyday life do not just have a practical or functional worth—they can also have more abstract meanings (Samida et al. 2014, 1). For example, a person's choice to use pen and paper or a computer to take notes can carry information about this person's private working culture and tell others something about them. The study of material culture helps in discovering these meanings, which can relate to ideas, attitudes, beliefs, and much more that describes communities or societies (Prown 1993, 1). As artifacts contain multifaceted information about the culture that produced them, one can explore the knowledge that is interwoven with the artifact itself, its invention, its production or its usage. This also being the case for historical artifacts, it is important to investigate the material culture of such an artifact in detail. A history of science and knowledge that focuses not just on the written word but also on things needs tools that make it possible to obtain all available information an artifact contains.

Volvelles, on the interface of artifact, image, and writing, can be explored with these tools too, to gain a more profound insight into their culture, function, and use.[1] As Anthony Drennan writes, volvelles require "more precision than normal descriptive methods" (Drennan 2012, 316). Though Drennan focuses on bibliographical descriptions of volvelles, his statement is also important for an investigation of the material culture of volvelles. As volvelles and their parts were printed at the same time and as a part of the rest of the book, in material and textual terms they are inseparable from the book (Drennan 2012, 318). But why are they inseparable? As we learned, books in the early modern period were mostly bound after purchase, sometimes together with works that were supplementing the content. The printing process, in contrast, was a closed circuit in which the pages and design of the pages had to be planned as a whole. The production of a book was a costly matter; the printer had to make sure that no material resources were wasted and thus he or she had to plan cautiously. Therefore, images or volvelles had to be placed thoughtfully in the book to fit with the layout while still being connected to the content. Even though the volvelles of the Wittenberg group do not appear at first sight to be clearly described or referred to in the text of the Sphaera, they are still connected to the content of the text. Drennan's statement that volvelles need a more precise description is especially interesting in connection to his statement that "A description that encompasses just the printed portions of the volvelle in their uncut state is of limited use, since without a description of the thread pointers, [...] and diagram construction, any textual analysis would in many cases be impossible" (Drennan 2012, 319). In order, Chapter 6, will describe and analyze the function and use of the volvelles, while describing their parts in detail.

As was stated in Chapter 4, not only the (correctly) assembled volvelle but also the individual parts or base discs can be sources of information. Keeping that in mind, the following Sects. 5.1 and 5.2 will describe the method for exploring the material culture of the volvelles.

5.1 Method: The Winterthur Model

Edward McClung Fleming's Winterthur Model is a common workflow used to investigate artifacts developed at the Winterthur Museum in Delaware. It examines five basic properties of artifacts (Fleming 1974, 156) by means of four operations that are performed on the five properties in order to answer most conceivable questions about the artifact. By applying this model to an artifact, a set of distinctive facts about it will be gained. The properties are as follows:

1. History

---

[1] Volvelles will often be referred to as "artifacts" in the following sections. The terms "object" and "artifact" are used synonymously. For more information on their equivalence, see for instance (Prown 1982, 2; 1993, 2; Pearce 1994, 9).

5.1 Method: The Winterthur Model

   a. where, when was it made?
   b. by whom, for whom?
   c. why was it made?
   d. changes in ownership, condition, function

2. Material
   a. what is the object made of?

3. Construction
   a. techniques of manufacture
   b. workmanship
   c. how are the parts organized to bring about the object's function?

4. Design
   a. structure
   b. form
   c. style
   d. ornament
   e. iconography

5. Function
   a. use (intended function)
   b. role (unintended function)

The four operations are as follows (Fig. 5.1) (Fleming 1974, 156–161):

1. Identification
   a. factual description
   b. provide accurate information about the five properties above

2. Evaluation
   a. comparison with other objects (size, rarity, temporal primacy …)
   b. apply adjectives such as similar, unique, early example …

3. Cultural analysis
   a. goes beyond description
   b. relation of the artifact to its culture: which function does the artifact have in its culture?
   c. reason for its initial manufacture
   d. identification with specific culture, subculture, geographic area…

4. Interpretation
   a. relation of the artifact to our culture

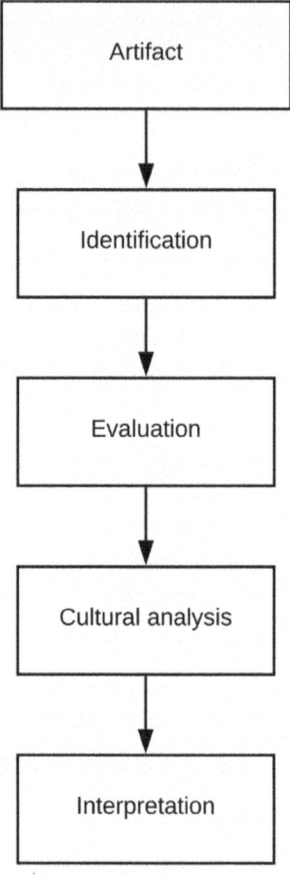

Fig. 5.1 Artifact model for analyzing objects. Modified illustration by the author after Fleming's model (Fleming 1974, 154). Plot by the author

The Winterthur model will be applied to one of each of the four volvelles of the Wittenberg group in Chapter 6. Some aspects of the model have already been addressed in previous sections, but these will be summarized below according to the workflow described by Fleming (Fig. 5.1). In this way the initial questions from Chapter 1 that are the focus of this work can be answered in the best possible way: Can the combination of text, image, and instrument offer other insights into premodern scientific knowledge and thought than just the artifact or just the text? Do the instruments or devices complement the knowledge that is given by the text and vice versa? What kind of knowledge do they provide, and what kind of function do they fulfill within books?

The Winterthur model was, as noted, developed in a museum, and is thus mainly used for investigating museal artifacts. As the function and usage of volvelles goes beyond the textual content of the book, this method is suitable for their investigation. The next section will introduce another concept of the interpretation of objects and their knowledge within the history of science.

5.2 Thing Knowledge

In his work *Thing Knowledge* Davis Baird argues against the idea that material products are mere manifestations of the epistemological theories they embody. He states that instruments, devices, and artifacts are epistemologically equivalent to theories while constituting knowledge that is different (Baird 2004, 1). He explains that both artifacts and theory can make up our knowledge of the world, while arguing in support of seeing instruments as a kind of scientific knowledge, as the history of science does (Baird 2004, 4, 11–12). Though Baird endorses the idea of the epistemological equivalence of theory and objects, he states that it should be kept in mind that there is no unified epistemological treatment for both. Not even every instrument works epistemologically in the same way: Different kinds of scientific object work differently in terms of the knowledge they constitute, which requires a special treatment for these various instruments (Baird 2004, 4–5, 12). To construct an epistemology that includes instruments, Baird's contribution categorizes instruments into three different groups:

- models
- instruments that create phenomena
- measuring instruments.

Models work similarly to theories while providing representations of complex systems or phenomena. As they are explanatory, predictive, and simple, they have the same advantages as theoretical explanations (Baird 2004, 12). According to Baird, they contain "model knowledge:" This type of knowledge is described as a knowledge that points to the theory (Baird 2004, 37). However, they are able to show further aspects than the mere theoretical explanations, or sometimes have an epistemological role prior to the existence of theory. **Instruments that create phenomena** constitute knowledge in a "different, nonrepresentational way" and can help in discovering unexpected new phenomena. Knowledge becomes visible as an effective action, separated from human agency, if the instrument works (Baird 2004, 12). This knowledge is called "working knowledge." The third group, **measuring instruments**, are a combination of the first two groups. They can be "as simple as a scale on a thermometer," and must work reliably while preserving accuracy (Baird 2004, 12). They also have to work as the instruments that create phenomena do. Additionally, they create the epistemological mode of model knowledge. This creates a special "encapsulated knowledge" (Baird 2004, 12).

Baird's ideas are neither revolutionary nor new to the history of science and knowledge. Still, his concept of different groups of instruments that constitute different forms of knowledge can be interesting for the analysis of volvelles. If the group of instruments to which they belong can be identified it could help with identifying the knowledge they contain. This will be part of the next chapter, in which all four of the volvelles from the Wittenberg group will be investigated with help of the Winterthur model.

Bibliography

Secondary Sources

Baird, Davis. 2004. *Thing knowledge*. London: University of California Press.
Drennan, Anthony S. 2012. The bibliographical description of astronomical volvelles and other moveable diagrams. *The Library* 13 (3): 316–339. https://doi.org/10.1093/library/13.3.316.
Fleming, Edward McClung. 1974. Artifact study: A proposed model. *Winterthur Portfolio* 9: 153–173.
Pearce, Susan M. 1994. Museum objects. In *Interpreting objects and collections*, ed. Susan M. Pearce, 9–11. London/New York: Routledge.
Prown, Jules David. 1982. Mind in matter: An introduction to material culture theory and method. *Winterthur Portfolio* 17 (1): 1–19.
Prown, Jules David. 1993. The truth of material culture: History or fiction? In *History from things: Essays on material culture*, ed. Steven Lubar and W. David Kingery, 1–19. London/Washington: Smithsonian Institution Press.
Samida, Stefanie, Manfred K.H. Eggert, and Hans Peter Hahn. 2014. *Handbuch Materielle Kultur. Bedeutungen, Konzepte, Disziplinen*. Stuttgart/Weimar: J.B. Metzler.

Open Access This chapter is licensed under the terms of the Creative Commons Attribution 4.0 International License (http://creativecommons.org/licenses/by/4.0/), which permits use, sharing, adaptation, distribution and reproduction in any medium or format, as long as you give appropriate credit to the original author(s) and the source, provide a link to the Creative Commons license and indicate if changes were made.

The images or other third party material in this chapter are included in the chapter's Creative Commons license, unless indicated otherwise in a credit line to the material. If material is not included in the chapter's Creative Commons license and your intended use is not permitted by statutory regulation or exceeds the permitted use, you will need to obtain permission directly from the copyright holder.

Chapter 6
The Function and Use of Volvelles from the Wittenberg Group

Abstract This chapter applies the previously introduced Winterthur model to the volvelles of the Wittenberg group of Johannes de Sacrobosco's *Sphaera*. Six exemplary copies that contain mostly intact versions of each of the volvelles were investigated following the workflow of the Winterthur model: identification, evaluation, cultural analysis, and interpretation. The Wittenberg group contains four volvelle types, the Eclipse, the Horizon, the Zodiac, and the Heliacal Volvelles. To each of the volvelles the model was applied, to examine them in the context of the books. Additionally, they are compared to their appearance in the six exemplary copies, to analyse the volvelles in every possible state (assembled, assembled incorrectly, and unassembled). Additionally, the text parts that relate to the volvelles within the editions were used in the investigation. In this way it was possible to explore the function and use of each of the four volvelles.

Keywords *Sphaera* · Johannes de Sacrobosco · Early modern period · History of science · History of knowledge · Volvelles · Paper instruments · Material culture · Winterthur model

Modern studies of volvelles and their history explain the functions volvelles could have had in early modern texts: They might show information or enrich a text by amplifying or directly relating to the words (Kanas 2005; Drennan 2012, 318); they might help in understanding and calculating complex phenomena, such as the sky and its movements; and they might have a mnemonic or pedagogical function (Gingerich 1993; Karr 2004). This latter function could mean that the volvelle was provided as a teaching aid intended to help the readers in interpreting the text (Drennan 2012, 318). In this way it was also possible to condense text and information onto a single leaf, and even to replace mathematics that was too difficult for the readers (Drennan 2012, 318). In addition, it is possible that in some cases the function of volvelles was simply to make the book more marketable (Drennan 2012, 318).

All or any of the above options could be possible of the volvelles from the *Sphaera* corpus. To identify the function and use of each of the four volvelles from the Wittenberg group, the volvelles and their surrounding text, along with the cultural context of the *Sphaera*, will be described and analyzed with the help of the Winterthur model.

As the structure, text and image groups of *Sphaera* editions containing volvelles is nearly identical, I will examine six different copies from the corpus, as listed in Table 6.1, that show the volvelles in different states and contexts.[1] The first four copies (Nos. 1–4) are from the Austrian National Library in Vienna (ÖNB). All of them were printed in Wittenberg in the sixteenth century, but by three different printers: Matthäus Welack (d. 1593), Johann Krafft the Elder (1510–1578), and Veit Kreutzer (d. 1578).[2] They are in good condition and contain assembled volvelles. No. 4, printed by Welack, belongs to the genre of *quaestiones* (Blebel 1576–1577).[3] This copy of the *quaestiones* edition regularly contains just one of the four volvelles in the Wittenberg group—the "heliacal volvelle." This volvelle type will be analyzed with the help of this copy of *Sphaera*, as the form of the *quaestiones* helps to understand the concept behind it.

The remaining copies (Nos. 5 and 6) were examined in the library of MPIWG, which holds copies of both of these editions for research purposes. The first copy (No. 5), printed by Jean Richard (1516–1573), contain all of the four volvelles in good condition and assembled correctly (Sacrobosco 1559).[4] In No. 6, by Symphorien Béraud (d. 1586), although the volvelles are also in excellent condition and assembled correctly, only three of the four are present, the heliacal volvelle being missing (Giuntini and Sacrobosco 1582).[5]

An essential part of the Winterthur model is a comparison with other objects concerning size, rarity, temporal primacy, etc. In this case, a broad base of identical artifacts exists. I will use this exceptionality, which is not common for early modern artifacts, to compare the volvelles already identified by the first step of the model to volvelles of other kinds in the second step. A comparison of the same volvelles from different printers in different European countries will give a good idea of the transfer and distribution of the volvelles, as well as that of the knowledge they

---

[1] The copies that will be examined are specific copies each connected to an edition from the corpus. Every copy can be identified via its source locator in the database, reachable under the Handle-Link from Table 6.1. Additionally, the bibliographical reference of each copy has a number in the Universal Short Title Catalogue (USTC; https://www.ustc.ac.uk).

[2] The database of the *Sphaera* project includes an overview on each of these printers. Matthäus Welack: http://hdl.handle.net/21.11103/sphaera.100684; Johann Krafft the Elder: http://hdl.handle.net/21.11103/sphaera.100955; Veit Kreutzer: http://hdl.handle.net/21.11103/sphaera.100805.

[3] *Quaestiones* was a characteristic pedagogical method of the early modern university and its scholastic teaching. In books this method was realized as follows: A question concerning the matter of the work was asked (e.g. *Quid est Sphaera recta?*/What is the right sphere?) and the brief printed answer followed directly (here: *Est positio Sphaerae, in qua uterque polus incumbit Horizonti.*/It is the position of the sphere in which each pole lies on the horizon).

[4] Jean Richard in the Sphaera Corpus*Tracer* database: http://hdl.handle.net/21.11103/sphaera.100348.

[5] Symphorien Béraud in the Sphaera Corpus*Tracer* database: http://hdl.handle.net/21.11103/sphaera.100396.

6 The Function and Use of Volvelles from the Wittenberg Group 65

Table 6.1 Bibliographical metadata of Sphaera copies that will be analyzed using the Winterthur model

Copy No.	Place	Printer	Year	Author(s)	Title	Library	Call No	Corpus*Tracer* Handle	USTC No
1	Wittenberg	Veit Kreutzer	1545	Melanchthon, Philipp Sacrobosco, Johannes de	Libellus de Sphaera	ÖNB	46.L.42	http://hdl.handle.net/21.11103/sphaera.100818	667491
2	Wittenberg	Johann Krafft the Elder	1558	Peucer, Kaspar	Elementa doctrinae de circulis coelestibus	ÖNB	72.N.56	http://hdl.handle.net/21.11103/sphaera.100191	649659
3	Wittenberg	Johann Krafft the Elder	1576	Peucer, Kaspar	Elementa doctrinae de circulis coelestibus	ÖNB	72.N.58	http://hdl.handle.net/21.11103/sphaera.100473	649661
4	Wittenberg	Matthäus Welack	1576	Blebel, Thomas	De sphaera et primis astronomiae rudimentis libellus	ÖNB	72.N.58*	http://hdl.handle.net/21.11103/sphaera.101306	631586
5	Antwerp	Jean Richard	1559	Sacrobosco, Johannes de	Sphaera Ioannis de Sacrobosco	MPIWG	Rara J655s	http://hdl.handle.net/21.11103/sphaera.101107	409110
6	Lyon	Symphorien Béraud	1582	Giuntini, Francesco Sacrobosco, Johannes de	La sfera del mondo	MPIWG	Rara G537s	http://hdl.handle.net/21.11103/sphaera.101118	130143

contain. Additionally, the examination of the six volvelles of the same kind will help to describe and identify any parts that might be missing.

The volvelles will be analyzed in the order in which they appear within the books. Since the Interpretation stage of the applied Winterthur model is identical for the first two volvelles (Eclipse and Horizon volvelle), this stage is summarized at the end of the analysis of the Horizon volvelle in Sect. 6.2.

6.1 Eclipse Volvelle

The volvelle I have termed "eclipse volvelle" appears in Chapter I of the *Sphaera* either under the heading *De Terra. I. Terram cum aqua globum constituere* (The earth. 1. The earth and water create a globe) or under *Quod terra & aqua sint globosa corpora, & mutuo complex unum globum unamque convexam superficiem constituant* (That the earth and water are sphaerical bodies and each complex form one globe and one convex surface together). This chapter of the *Sphaera* deals with the parts of the sphere and the place of the earth inside the system of that time.

In the copies No. 1 by Kreutzer (1545), No. 3 by Krafft the Elder (1576), No. 5 by Richard (1559), and No. 6 by Béraud (1582) this volvelle is correctly assembled and in good condition. In Krafft the Elder's print from 1558 (No. 2) the volvelle was not installed and in No. 4 by Welack (1576) it was not even printed.

The Winterthur model will now be applied to this volvelle by first describing it factually while keeping in mind the previously described artifactual properties (history, material, construction, design, and function), after which the steps "Evaluation," "Cultural analysis," and "Interpretation" will be performed. The remaining three volvelles will then be examined in the same manner.

6.1.1 Identification

This volvelle is a set of wheels and pointers made from paper and found inside a book from the previously described *Sphaera* corpus. It consists of four layers: a base disc (Fig. 6.1), which forms part of a page of the book (a on Fig. 6.2), a moveable round disc with pointers (b), a layer that consists of a triad of pointers that are moveable as a whole (c) and the last layer, an immoveable cap with an additional pointer (d). All the pieces are printed with black ink and are usually not colored.[6]

The base disc (a) consists of a circle divided into quarters. Each quarter is marked by a line and by a letter on the outside of the circle. Clockwise, these are "B" at 0/360°, "C" at 90°, "D" at 180° and "A" at 270°. Additionally, the quarters are each divided into six parts that are marked black and white. Each of these six parts would

---

[6] Rare cases exist in which images within *Sphaera* editions or copies were colored—either before being sold or colored by the reader.

6.1 Eclipse Volvelle

Fig. 6.1 Page containing a base disc of an eclipse volvelle. From Peucer (1558, F6r). Augsburg Staats- und Stadtbibliothek, Math 900#2, urn:nbn:de:bvb:12-bsb112 67743-1

have 15° of a circle. The fourth quarter of the circle ("A"–"B") is marked twice: at 30° with "L" and at 45° with "M" (Fig. 6.1).

The disc with pointers (b) is circular and slightly smaller than the base disc. It has four opposing pointers on the outside, of which one connects directly to the inner image of the circle. The image shows a sun with a face that looks down on a circle in the middle of the disc. In line with the sun and the middle circle, a third figure follows in the middle of a black triangle, which probably represents the earth's shadow. The third figure, which consists of a circle with a face, is a moon.

The triad of pointers (c) consists of two opposing pointers marked with the letters "F" and "H," and a third one ("G") that stands vertically on the line formed by the two opposing pointers.

Part (d) serves as a cap for the volvelle and remains fixed. Inside the circle an image is usually drawn that shows a landscape. Here it is a castle in a mountainous region. Attached to this cap is another pointer, marked with a "B."

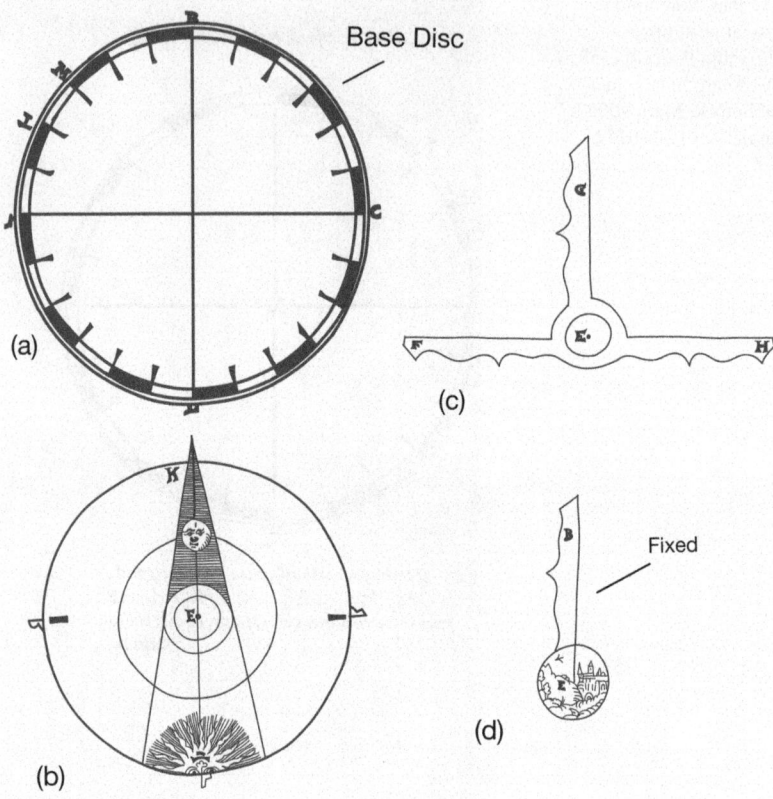

Fig. 6.2 Drawing of the four parts of an eclipse volvelle. Illustration by the author

The parts are connected with one another either by a thread or by a pin made out of paper. In the majority of cases the page on which the base disc was printed is not damaged but shows fine pressure marks on the reverse where the pin is glued to the page (Fig. 6.3).

As Jürgen Hamel states in his work, the volvelles in the *Sphaera* corpus were designed with the help of Petrus Apianus, although the *Sphaera* edition that was printed by him in 1526 lacked volvelles (Apianus and Sacrobosco 1526; Hamel 2014, 42). Anthony Drennan claims for volvelles in general, that were usually designed by the author of the corresponding book (Drennan 2012). This is probably not the case for the volvelles of the Wittenberg group: First, all of these are nearly identical, though they were printed in different editions of the Sphaera, that were commented and edited by different commentators. Additionally, most of the volvelles are the same size throughout the six copies examined: The base discs from Nos. 1–5 are all 77 mm in diameter. This might suggest that the volvelles were distributed via the printers: As the woodblocks were nearly the same size it is possible that the printers used the

6.1 Eclipse Volvelle

Fig. 6.3 Back of a page that contains a volvelle, showing the pressure mark of the mounting method. From Melanchthon and Sacrobosco (1545). Austrian National Library, 46.L.42

same woodblocks and did not just copy the diagrams visually. This practice was quite common, not just in Wittenberg but also in, for example, Venice, as Saskia Limbach and Catherine Rideau-Kikuchi have demonstrated (Limbach 2022; Rideau-Kikuchi 2022). Limbach in particular showed how several printers in Wittenberg during this period used the same woodblocks for editions of the *Sphaera*.

The eclipse volvelle was usually printed in parts. The base disc was printed inside the book where it was intended to be assembled by the reader. The parts were printed on a folio sheet that was bound in the back of the book and folded. For use, the parts had to be cut out by the reader and assembled inside the work at the intended page. For a correct installation, the parts had to be put together as shown in Fig. 6.4.

The intended use of this volvelle was probably to demonstrate how an eclipse can be seen with respect to the horizon. It shows that for observers at different longitudes the eclipse appears at different times of the day, which demonstrates the roundness of the earth in a longitudinal (east–west) sense. The location of the volvelle inside the *Sphaera* copies proves this hypothesis, as the volvelle is printed after text stating that the earth is round:

> That the earth, too, is round is shown thus. The signs and stars do not rise and set the same for all men everywhere but rise and set sooner for those in the east than for those in the west; and of this there is no other cause than the bulge of the earth. Moreover, celestial phenomena evidence that they rise sooner for orientals than for westerners. For one and the same eclipse of the moon which appears to us in the first hour of the night appears to orientals about the third hour of the night, which proves that they had night and sunset before we did, of which setting the bulge of the earth is the cause. (Thorndike 1949, 121)[7]

---

[7] Thorndike (1949, 81–82): "Quod terra etiam sit rotunda sic patet. Signa et stelle non equaliter oriuntur et occidunt omnibus hominibus ubique existentibus sed prius oriuntur et occidunt illis qui sunt iuxta orientem quam illis qui sunt iuxta occidentem, et huius nulla alia causa est nisi tumor terre. Quod autem orientalibus citius oriantur quam occidentalibus bene patet per ea que fiunt in sublimi. Una enim et eadem eclipsis lune numero que apparet nobis in prima hora noctis apparet orientalibus circa horam noctis tertiam. Uncle constat quod illis fuit prius nox, et sol prius occidit eis quam nobis, cuius occasus causa est tumor terre."

Fig. 6.4 Assembly of the eclipse volvelle. Illustration by the author

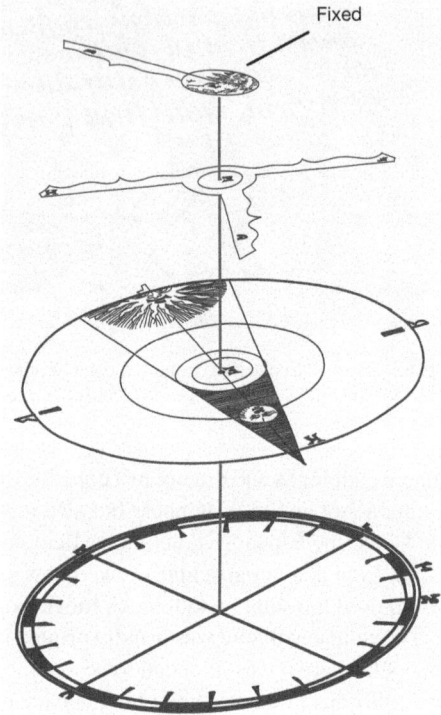

6.1.2 Evaluation

Eclipse volvelles often look quite similar, if not identical. As noted, 89% of the volvelles from the *Sphaera* corpus are from the Wittenberg group, of which this volvelle is characteristic. The description above applies to all of the assembled eclipse volvelles from the *Sphaera* corpus; only the illustration on the fixed cap differs in some cases. Although this volvelle is typical for the *Sphaera*, it seems to be unique in the corpus of early modern astronomical books, as it has not been found in another book yet.

6.1.3 Cultural Analysis

The sixteenth and seventeenth centuries in particular were, as noted above, a "golden age" of volvelle production, in which volvelles made books more marketable. As books were accessible to only a small proportion of the population, volvelles were clearly used only by this group. Nevertheless, the *Sphaera* corpus alone consists of 123 editions containing moveable parts, which must have been around 123,000 copies. As this volvelle is found in a book that was primarily addressed to students in Europe, it is likely that it is intended to work as a teaching aid, with a didactic or mnemonical function. It might have initially been designed because professors noted that students had problems with imagining the movements of the heavenly objects. As armillary spheres were much more expensive than books, demonstrations of certain heavenly movements with the help of volvelles could have made the subject's study easier and more accessible.[8]

In the sixteenth and seventeenth centuries volvelles appeared in many books and for different reasons (lottery, fortune telling, scientific demonstrations). Suzanne Karr Schmidt has stated that at least 190,000 copies of books with moveable parts must have existed in the sixteenth century. This number is probably higher, as even the *Sphaera* alone numbers 98 editions and as many as 98,000 copies. This means that books with moveable parts, including volvelles, were widespread within the book-accessing community. This community was probably also divided between people with the funds to buy books with more elaborated volvelles, such as the *Astronomicum Caesareum* by Petrus Apianus, and less well-off people like students, who might sometimes have to buy books second-hand.[9]

6.2 Horizon Volvelle

The second volvelle, called the "horizon volvelle," is also located in Chapter I. It is one of two of the four volvelles that is mentioned in the text and referred to as an "instrument": *Instrumentum quo & rotunditas terrae secundum latitudinem probari, & facilime omnia ca, quae Autor in tertio capite de diebus artificialibus tradit, ostendi possunt* (The instrument by which the roundness and the latitude of the earth can be proved and all the works which the author reports in the third chapter about artificial days can be shown very easily). In the copies No. 1 by Kreutzer (1545), No. 2 and No. 3 by Krafft the Elder (1558, 1576), No. 5 by Richard (1559), and No. 6 by Béraud (1582) this volvelle is correctly assembled and in good condition. In No. 4 by Welack (1576) it was not even printed.

6.2.1 Identification

This volvelle consists of three parts: the base disc (a), the second disc (b) and the cap that holds the discs in place (c) (Fig. 6.5). All parts are printed with black ink on paper and are usually not colored by production. The base disc is the most complex of all four volvelles from the Wittenberg group. It is circular in shape and divided

---

[8] Armillary spheres are models of the heaven that demonstrate the movements of the planets with the help of rings. Ptolemaic armillary spheres have the earth at their center.

[9] As stated above, the Interpretation for this volvelle is summarized in Sect. 6.2.

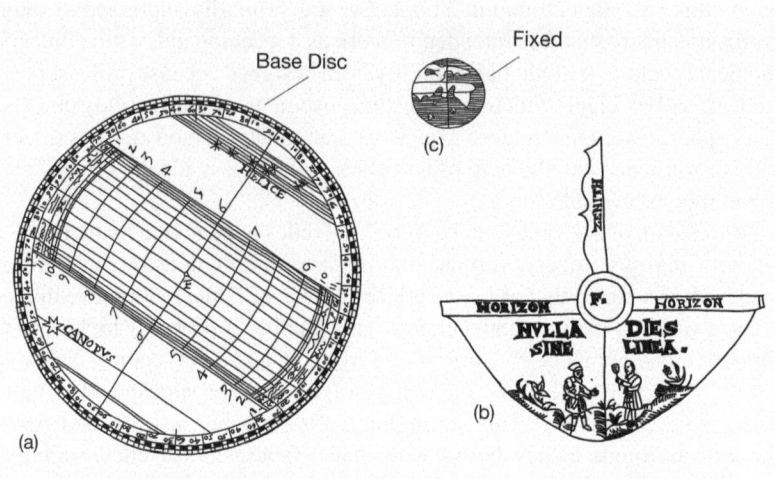

Fig. 6.5 Drawing of the three layers of a horizon volvelle. Illustration by the author

into four quarters. Each quarter is subdivided in steps of 10°, which are labelled according to their degree and are additionally subdivided into five pieces that equal 2° each.

Within the base disc "Canopus" is written at bottom left and "Helice" at top right. Canopus is the brightest star of the—now obsolete—star constellation Argo Navis from Ptolemy's forty-eight constellations and it is still the second brightest star in the night sky (Peters and Knobel 1915, 16). As Canopus was visible not from Greece but from further south, in Egypt, it was called the "bright star of the Egyptians" (Evans 1998, 48). It also appears in Ptolemy's list of stars found in the *Almagest* (Peters and Knobel 1915, 93). "Helice" describes a star constellation now known as Ursa Major or the Great Bear (*Encyclopedia Britannica* 2020). The name Helice originates in ancient times and describes the constellation's turning around the Pole Star. It is, like the Argo Navis, catalogued by Ptolemy (Peters and Knobel 1915, 16).

The middle section of the base disc is occupied by a grid similar to a three-dimensional cylinder. Eleven lines cross the grid perpendicular to the main axis and are numbered from 1 to 11 according to their position. The numbers indicate latitudes, but beginning not from the equator but from the ecliptic (ecliptic latitudes), thus following the ecliptic coordinate system. They start at line number 1 (or 11), which equals to 0° at the ecliptic, and continue in steps of 15°; the latitude lines 1 and 11 thus equal the same latitude line. It becomes clear that the grid in the middle of the base disc is meant to be double layered. To visualize its function, it can be imagined as wrapped around the earth with the point F lying on both of the ecliptic poles. Both ends of the grid hold two lines of six boxes each. Inside the boxes, the twelve signs of the zodiac are represented in pairs and by their symbols. The order is as follows and shows that opposite signs are paired:

6.2 Horizon Volvelle

Cancer	Leo	Virgo	Libra	Scorpio	Saggitarius
Gemini	Taurus	Aries	Pisces	Aquarius	Capricorn

The second disc (b) is meant to be attached to the base disc. The disc (b) is T-shaped, consisting of a horizon and a vertical zenith that originates in the middle of the disc. A semicircle is described below the horizontal line, which is illustrated with two people standing in a landscape of flowers. One of them holds a carnival mask in his hand. Above their heads the words "NULLA DIES SINE LINEA" ("No day without a line") can be read. This piece is meant to show the paths of the sun on various days.

The cap (c) holds the second disc in place. For assembly, the parts had to be cut out by the reader and assembled inside the work at their intended page. For a correct installation, the parts had to be put together as shown in Fig. 6.6.

This presentation of zodiac signs and the sun, in addition to the grid, aims to show the earth's curvature in a latitudinal way. Sacrobosco's text describes this as follows:

> That the earth also has a bulge from north to south and vice versa is shown thus: To those living toward the north, certain stars are always visible, namely, those near the North Pole,

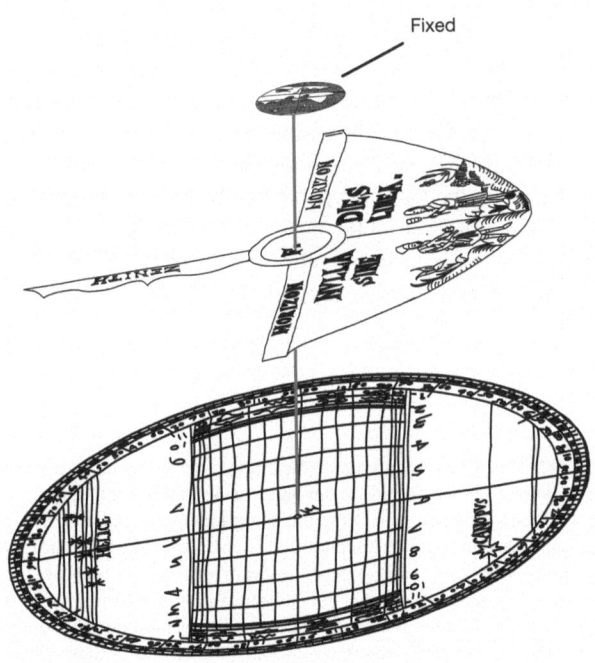

Fig. 6.6 Assembly of the horizon volvelle. Illustration by the author

Fig. 6.7 Diagram of a ship sailing on the spherical surface of the earth. From Melanchthon and Sacrobosco (1545, B8v). BSB München, Astr. U 158, urn:nbn:de:bvb:12-bsb101 73685-3

while others which are near the South Pole are always concealed from them. If, then, anyone should proceed from the north southward, he might go so far that the stars which formerly were always visible to him now would tend toward their setting. And the farther south he went, the more they would be moved toward their setting. Again, that same man now could see stars which formerly had always been hidden from him. And the reverse would happen to anyone going from the south northward. The cause of this is simply the bulge of the earth. Again, if the earth were flat from east to west, the stars would rise as soon for westerners as for orientals which is false. Also, if the earth were flat from north to south and vice versa, the stars which were always visible to anyone would continue to be so wherever he went, which is false. But it seems flat to human sight because it is so extensive.[10]

The horizon volvelle and the accompanying text relate to another illustration that demonstrates the spherical shape of the earth (Fig. 6.7). To do so, it illustrates a ship sailing towards a building on the surface of the earth. Two lines that link the ship and building demonstrate how the latter is visible from the mast of the ship first "before becoming visible for an observer on the boat" (Kräutli et al. 2020).

The combination of the eclipse volvelle, the horizon volvelle, the text corresponding to each and the ship diagram supported readers to understand the spherical shape of the earth.

---

[10] Thorndike (1949, 82–83): "Quod terra etiam habeat tumorem a septentrione in austrum et econverso sic patet. Existentibus versus septentrionem quedem stelle sunt sempiterne apparitionis, scilicet que accedunt ad polum articum. Alie vero sunt sempiterne occultationis, que sunt propinque polo antartico. Si igitur aliquis procederet a septentrione versus austrum, in tantum posset procedere quod stelle, que prius errant ei sempiterne apparitionis, iam tenderent in occasum. Et quanto magis accederet ad austrum, tanto plus moverentur in occasum. Ille item idem homo iam posset videre stellas que prius fuerant illi sempiterne occultationis. Et econverso contingeret alicui procedenti ab austro versus septentrionem. Huius autem rei causa est tantum tumor terre. Item si terra esset plana ab oriente in occidentem, tam cito orirentur stelle occidentalibus quam orientalibus, quod falsum est. Si terra etiam esset plana a septentrione in austrum et econverso, stelle que essent alicui sempiterne apparationis semper apparerent ei quocunque procederet, quod falsum est. Sed quod plana sit pre nimia eius quantitate visui hominum apparet."

6.2.2 Evaluation

As with the previous eclipse volvelle, volvelles of the category horizon volvelle often look quite similar, if not identical. About 89% of the volvelles from the *Sphaera* corpus are from the Wittenberg group, of which this volvelle is characteristic. The description above applies to all the parts of the assembled horizon volvelles from the *Sphaera* corpus, except the illustration on the fixed cap, which differs in some cases.

This volvelle is typical for the *Sphaera* corpus. Similar but simpler versions can be found in the Seville group (Fig. 4.7a) and the Leiden group (Fig. 4.8).

6.2.3 Cultural Analysis

As with the previously described eclipse volvelles, this volvelle is probably meant to work as a teaching aid for students in the early modern university. It might have initially been designed because professors found that students had problems with imagining the spherical shape of the earth and a visual aid could have made it easier to understand this.

6.2.4 Interpretation

This subsection deals with the relation of the artifact to our present culture. Neither the eclipse nor the horizon volvelle type can be found today and neither have any function in the modern world. Nevertheless, volvelles as a general type are used in various contexts in today's culture, but in different ways than in the early modern period: Hobby astronomers, for example, use planetary discs to know which star constellations can be seen on the night sky at a certain time of the year. Today, however, volvelles are primarily found in children's books, where they allow the reader to engage more deeply with the book's content, helping children to enjoy the story and to visualize various storylines, and to engage more with books in general.

6.3 Zodiac Volvelle

This volvelle appears in Chapter II of the *Sphaera* editions under the heading *De zodiaco circulo* (About the Circle of the Zodiac). It is correctly assembled and in good condition in the copies Nos. 1, 2, 3, 5, and 6, while it is—as with the previously described volvelles—not even printed in No. 4.

6.3.1 Identification

The zodiac volvelle contains of two parts: a base disc (a) and a pointer (b). Both are printed with black ink on paper and are usually not colored.

The base disc is circular in shape and divided into twelve parts of 30° each, each of which has marks on the outside that divides it into three parts of 10° (Fig. 6.8). Each of the twelve parts contains a symbol that stands for one of the twelve zodiac signs; these are ordered counterclockwise, starting with "Aries" (Fig. 6.9).

The pointer (b) is meant to be fixed in the middle of the base disc (Fig. 6.10). It shows a sun that rotates through the zodiac signs.

With the help of this volvelle students were probably meant to imagine the daily motion of the sun through the zodiac signs. The adjacent text, reproduced below, supports this suggestion:

> The names, order, and number of the signs are set forth in these lines: There are Aries, Taurus, Gemini, Cancer, Leo, Virgo, Libra and Scorpio, Architenens, Caper, Amphora, Pisces.
>
> Moreover, each sign is divided into 30 degrees, whence it is clear that in the entire zodiac there are 360 degrees. Also, according to astronomers, each degree is divided into 60 minutes, each minute into 60 seconds, each second into 60 thirds, and so on. And as the zodiac is divided by astronomers, so each circle in the sphere, whether great or small, is divided into similar parts. While every circle in the sphere except the zodiac is understood to be a line or circumference, the zodiac alone is understood to be a surface 12 degrees wide of degrees such as we have just mentioned. Wherefore, it is clear that certain persons in astrology lie who say that the signs are squares, unless they misuse this term and consider square and

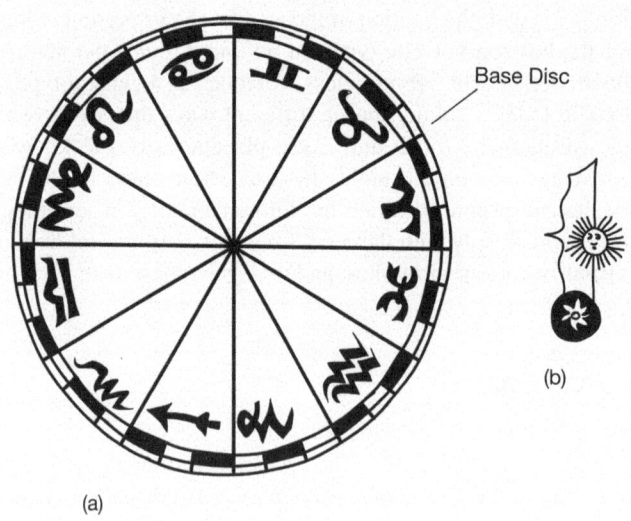

Fig. 6.8 Base disc and pointer of the zodiac volvelle. Illustration by the author

6.3 Zodiac Volvelle

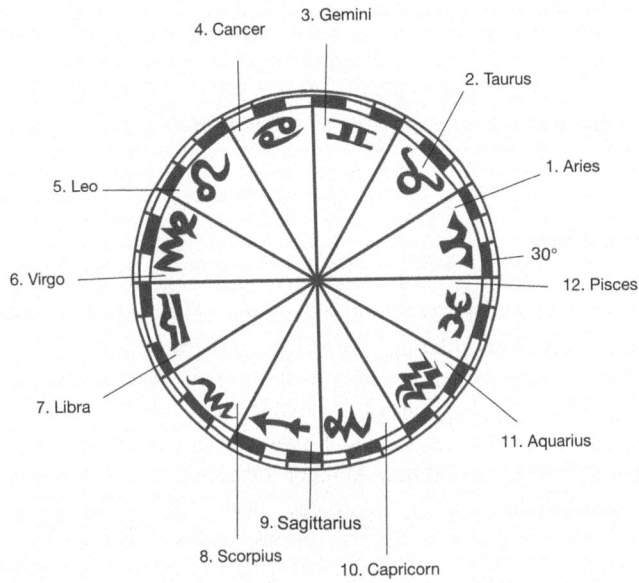

Fig. 6.9 Base disc of the zodiac volvelle. Illustration by the author

Fig. 6.10 Assembly of the zodiac volvelle. Illustration by the author

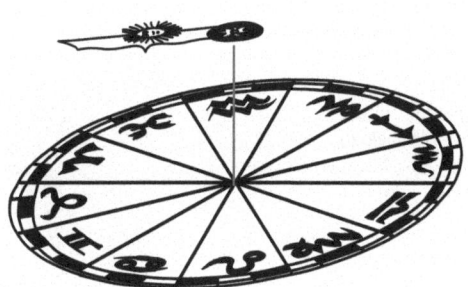

quadrangle the same. For each sign is 30 degrees in longitude, 12 in latitude. (Thorndike 1949, 124–125)[11]

---

[11] Thorndike (1949, 88): "Nomina autem signorum, ordinatio et numerus in hiis patent versibus: Sunt Aries Taurus Gemini Cancer Leo Virgo Libraque Scorpius Architenens Caper Amphora Pisces. Quodlibet autem signum dividitur in 30 gradus, unde patet quod in toto zodiaco sunt 360 gradus. Secundum astronomos iterum quilibet gradus dividitur in 60 minuta, quodlibet minutum in 60 secunda, quodlibet secundum in 60 tertia, et sic deinceps. Et sicut dividitur zodiacus ab astronomo, ita et quilibet circulus in spera, sive maior sive minor, dividitur in partes consimiles. Cum etiam omnis circulus in spera preter zodiacum intelligitur sicut linea vel circumferentia, solus zodiacus intelligitur superficies habens in latitudine 12 gradus, de cuiusmodi gradibus iam locuti sumus. Unde

This describes the properties of the zodiac in the sky: According to the text, it is a belt-shaped region (here: *surface*) of the sky which not just consists of a circular line but is 12° wide. Each sign occupies a sector of 30° of longitude and 12° of latitude. As the zodiac lies along the ecliptic, the volvelle can show how the sun moves through the signs throughout the year by moving the pointer through the signs.

6.3.2 Evaluation

The description given in Sect. 6.3.1 applies to all the zodiac volvelles and parts from the corpus. This volvelle is, like the eclipse and horizon volvelles, a paper wheel characteristic of the *Sphaera* corpus, but, in addition, its general appearance is also quite typical for other visualizations of the zodiac. For example, in Petrus Apianus' *Astronomicum Caesareum* an astrological atlas exists that pictures star constellations in a spherical shape. In addition, in other *Sphaera* editions—even handwritten manuscripts—similar diagrams of the zodiac can be found: The Staatsbibliothek Berlin contains a *Sphaera* manuscript by Konrad von Megenberg (1309–1374) that pictures the zodiac in a comparable manner (Fig. 6.11), and similar can be found in a *Sphaera* manuscript from ca. 1385 with Johannes de Sacrobosco as the credited author (Fig. 6.12). These suggest that this kind of circular diagram presenting the signs of the zodiac was quite common in late medieval and early modern Europe.

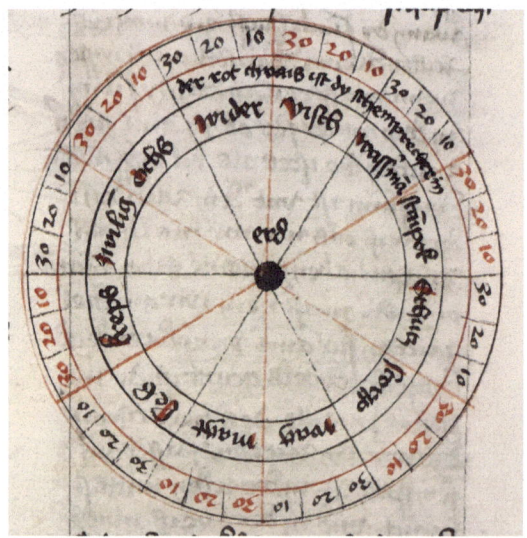

Fig. 6.11 Zodiac wheel from Konrad von Megenberg's *Sphaera* manuscript. Berlin, Staatsbibliothek zu Berlin—PK, Ms. Germ. Fol. 1069, fol. 226r

---

patet quod quidam mentiuntur in astrologia dicentes signa esse quadrata, nisi abutentes nomine idem appellant quadratum et quadrangulum. Signum enim habet 30 gradus in longitudine, 12 vero in latitudine."

6.3 Zodiac Volvelle

Fig. 6.12 Zodiac wheel from Johannes de Sacroboco's *Sphaera mundi* (1385). Berlin, Staatsbibliothek zu Berlin—PK, Ms. Germ. Fol. 479, fol. 15r

6.3.3 Cultural Analysis

The descriptions given in Sects. 6.1.3 and 6.2.3 apply also to the zodiac volvelle: It belongs to a corpus of books that shaped the scientific and cultural identity of Europe through the early modern period, to which the volvelles seem to have made a major contribution, as around 34% of the books from the corpus contain volvelles.

This volvelle seemed to have a mnemonic function, as it visualizes the heavenly movement of the sun in quite a simple way. Particularly in the case of this volvelle it is important to keep in mind the "picture superiority effect" (Defeyter et al. 2009) mentioned above: The brain processes images much more easily than words. This effect is stronger if the images move, as recent studies have shown (Rosen 2017; Schurgin and Flombaum 2017). In fact, this effect is probably important for all of the volvelles from the *Sphaera* corpus, as the books targeted students learning about the basic heavenly movements. With the help of moveable wheels they were probably able to memorize the celestial movements more easily, as well as the order of the signs of the zodiac.

6.3.4 Interpretation

This section deals with the relation of the artifact to our present culture. In Jessica Helfand's work "Reinventing the Wheel" she collected modern volvelles, some of which are quite similar to this zodiac volvelle (Helfand 2002). Among them are an "Astrologers Movable Planisphere" from 1909 and an "Art Deco Horoscope Wheel" from 1932 (Helfand 2002, 42–43). The titles of the wheels as well as their appearance show that in modern times the signs of the zodiac are mainly important for astrological purposes.

6.4 Heliacal Volvelle

The fourth and last volvelle—the "heliacal volvelle"—of the Wittenberg group appears in Chapter III of the *Sphaera* editions under the heading *Instrumentum, quo facilime omnes diversitates ortus Poetici, oculis subijciuntur* ("The instrument, by which all different poetical risings are brought to mind"). As previously described, this volvelle was assembled least often of those investigated here, probably because the instructions were not completely clear or because the underlying astronomy was too complex. In the six investigated copies it is assembled in just two of them: Nos. 3 and 5. Copies Nos. 1, 2, and 4 contain the base discs, but not an assembled volvelle, while the volvelle was probably assembled in No. 1 but taken apart afterwards. Copy No. 4, as noted above, lacked any of the previous volvelles, but the base disc of this heliacal volvelle was printed. Copy No. 6 contains neither the assembled volvelle nor the base disc. It is the latest of the investigated copies, printed in 1582, which suggests the possibility that the printer or author decided to cut this volvelle from the book because, based on experience, it was assembled and used less often than the other volvelles.

6.4.1 Identification

This volvelle, like the eclipse volvelle, is made of four parts: A base disc (a), two moveable discs (b) and (c) and a fixed cap (d) that holds the other parts in place (Fig. 6.13). As with the previous volvelles, the parts are printed on paper with black ink.

The base disc (a) is a circle with a double outer line. It is divided into four quarters by simple lines crossing in the middle of the circle. On the outside of the circle four points are marked with words, starting from 12 o'clock and proceeding clockwise:

- MERI.: stands for meridian and marks the middle of the day or the "Mittagspunkt" or "Südpunkt," which is the crossing of the meridian with the horizon (Meyers Großes Konversationslexikon 1908, 914);

6.4 Heliacal Volvelle

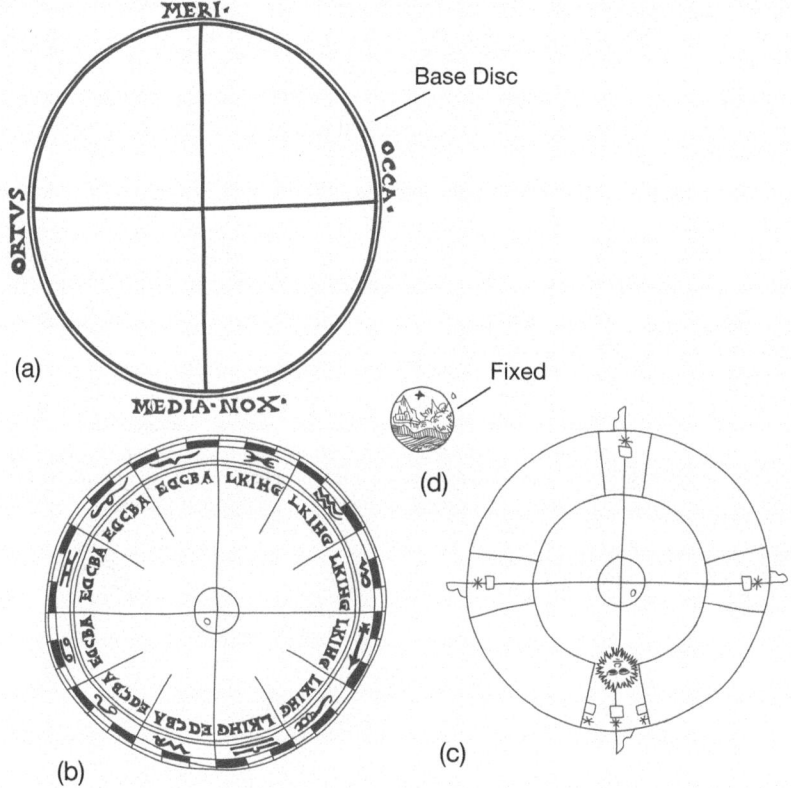

Fig. 6.13 The four components of the heliacal volvelle. Illustration by the author

- OCCA.: *lat. Occasus*, the setting of the stars and planets, and can also be understood as the "West" (Georges 1918a);
- MEDIA.NOX.: *lat.* the middle of the night;
- ORTUS: *lat.*, the rising of the stars and planets, the East (Georges 1918b).

In the case of this base disc, West and East are interchanged, as the West (*occasus*) is on the right and East (*ortus*) on the left.

The second disc (b) is also circular. The outer circle follows the design of the eclipse volvelle, marking 10° steps of the circle with an alternating black and white pattern. Every section of 30° contains one of the zodiac signs in their regular order. Within the circle of zodiac signs every 30° section contains a combination of letters: On the left half of the circle each section contains the letters EDCBA and on the right half of the circle every section is marked by LKIHG. These have to be read counterclockwise, as West and East are interchanged in this case.

The third disc (c) is smaller, but also circular. It has four pointers on the ends of two axes. The axes cross a smaller, inner circle and the middle of both of the circles. Between the outer and the inner circle and in line with the axes there are four small squares marked with a simplified drawing of a star, which are meant to be cut out. If cut out, the squares align with the letters on disc (b). One of the squares is accompanied by two further squares that will show, in total, three letters when cut out (at the bottom on Fig. 6.14). On this axis and just inside them, a sun is drawn on the inner circle.

The cap (d) holds all the discs in place and is, as before, decorated with a drawing of a landscape.

As the function of this volvelle is not obvious at first sight, it is necessary to look to the surrounding text for clues. The subject of Chapter III is the risings and settings of the zodiacal signs, and the text treats the "cosmic," "chronic," and "heliacal rising and settings" in the following manner:

> Cosmic or mundane rising takes place when a sign or star ascends above the horizon from the east by day. And, albeit in each artificial day six signs rise, yet antonomastically that sign is said to rise cosmically with which and in which the sun rises in the morning. And this is rising in the strict and chief and daily sense. Of this rising we have an instance in the

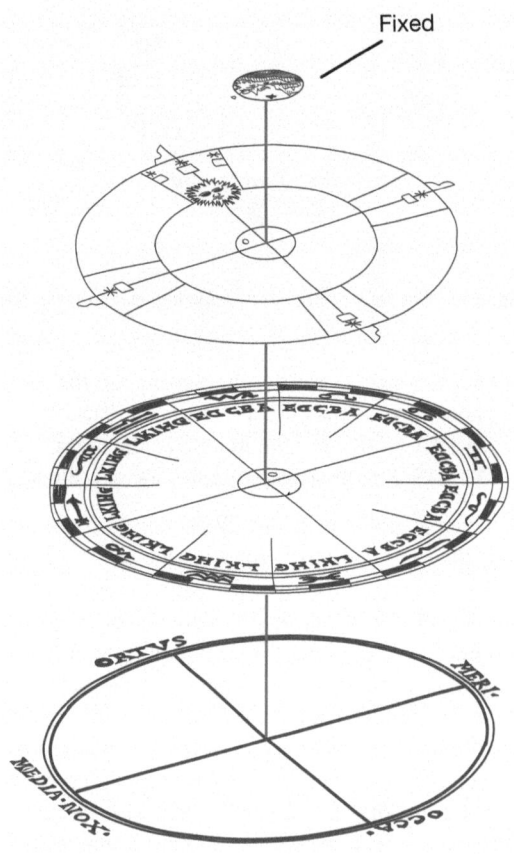

Fig. 6.14 Assembly of the heliacal volvelle. Illustration by the author

6.4 Heliacal Volvelle

Georgics, where the planting of beans and millet in springtime, when the sun is in Taurus, is taught thus: The white bull with gilded horns opens the year, And the dog, yielding to adverse star, sets.

Cosmic setting is a matter of opposition. When the sun rises with a sign, the opposite sign sets cosmically. This setting is spoken of in the *Georgics*, where is taught the sowing of wheat in late fall when the sun is in Scorpio. For when Scorpio rises with the sun, Taurus, where the Pleiades are, sets: Ere Eoe Atlantides are hidden from you, Consign the seed as you should to the furrow.

Chronic or temporal rising takes place when sign or star, after sunset, emerges above the horizon from the east at night. It is called "temporal rising" because astronomical time begins with sunset. Of this rising we have an example in Ovid's *From Pontus*, where he complains of his prolonged exile, saying, Pleias having risen makes four autumns, signifying by four autumns that four years had passed since he was sent into exile. But Virgil made the Pleiades set in the autumn, so they seem contradictory. But the explanation of this is that according to Ovid they rise chronically, which may well happen on the same day but differently, since cosmic setting is with respect to daytime, but chronic rising is in the evening. Chronic setting is a matter of opposition. Hence Lucan: Then the short night shot Thessalian arrows.

Heliacal or solar rising occurs when sign or star can be seen by departure of the sun from it, which previously could not be seen because of the nearness of the sun. Ovid gives an example of this in the *Fasti*, saying: Now aged Aquarius sits below with urn inclined. And Virgil in the *Georgics*: And the Gnosian star of the burning crown descends, which star, being next to Scorpio, was not visible while the sun was in Scorpio. Heliacal setting takes place when the sun approaches a sign and by its presence prevents it from being seen. An example of this is the following verse: And the dog, yielding to adverse star, sets. (Thorndike 1949, 129–131)[12]

---

[12] Thorndike (1949, 95–97): "Cosmicus enim ortus sive mundanus est quando signum vel stella supra orizontem ex parte orientis de die ascendit. Et licet in qualibet die artificiali sex signa oriantur, tamen antonomastice signum illud dicitur cosmice oriri cum quo et in quo sol mane oritur. Et hic ortus proprius et principalis et cotidianus dicitur. De hoc ortu habemus exemplum *in Georgicis* ubi docetur satio fabarum et milii in vere sole existente in Tauro sic: Candidus auratis aperit cum cornibus annum Taurus, et adverso cedens canis occitdit astro. Occasus vero cosmicus est ratione oppositionis, quando sol oritur cum aloiquo signo cuius signi oppositum occidit cosmice. De hoc occasu dicitur *in Georgicis*, ubi docetur satio frumenti in fine autumpni sole existente in Scorpione, qui cum oritur cum sole, Taurus ubi sunt Pleiades occidit: Ante tibi Eoe Atlantides abscondantur Debite quam sulcis committas semina.... Cronicus ortus sive temporalis est quando signum vel stella post solis occasum supra orizontem ex parte orientis emergit de nocte, et dicitur temporalis ortus quia tempus mathematicorum nascitur cum solis occasu. De hoc ortu habemus exemplum in Ovidio *de Ponto,* ubi conqueritur moram exilii sui dicens: Quatuor autumpnos Pleias orta facit, significans per quatuor autumpnos quatuor annos transisse postquam missus erat in exilium. Sed Virgilius voluit in autumpno Pleiades occidere, ergo contrarii videntur. Sed ratio huius est quod secundum Virgilium occidunt cosmice, secundum Ovidium oriuntur cronice, quod bene potest contingere in eodem die sed differenter tamen, quia cosmicus occasus est respectu temporis matutini, cronicus vero ortus respectu vespertini. Cronicus occasus est ratione oppositionis unde Lucanus: Tunc nox Thessalicas urgebat parva sagittas. Eliacus ortus sive solaris est quando signum vel stella potest videri per elongationem solis ab illa que prius videri non poterat solis propinquitate. Exemplum huius point Ovidius *in Fastis* sic: Iam senis obliqua subsedit Aquaris urna. Et Virgilius *in Georgicis*: Gnosiaque ardentis descendit stella corone, que iuxta Scorpionem existens non videbatur dum sol erat in Scorpione. Occasus eliacus est quando sol ad signum accredit et illud sua presential videri non permittit. Huius exemplum es in versu premisso: Taurus, et adverso cedens canis occidit astro."

These text extracts engage with a concept that refers to *De ortu poetico*, which was often misinterpreted as a collection of ancient texts with a connection to astronomy. As Valleriani et al. (2022, 423) have shown, *De ortu poetico* is

> a sort of tool to enable historians to exactly date events that took place in the past and for which there are testimonies that described the position of specific stars over the ecliptic in order to record the time. *De ortu* is a calendric means of investigating the past that also takes into consideration the geographical area from which the stars were observed.

Sacrobosco here connects the agricultural year to recurring celestial occurrences. To illustrate his point, he draws on the work of ancient poets such as Virgil (70–19 BCE), Ovid (43 BCE–CE 17), and Lucan (CE 39–65). Virgil's *Georgics* belongs—as does the *Fasti* by Ovid—to the genre of didactic poetry. It covers the topics of farming, tree planting, animal husbandry, and bee care, which are directly connected to risings and settings. Sacrobosco connects the quotes he cites directly to cosmic settings, which serves to give information about the time of the year. The quotation "the white bull with gilded horns opens the year, And the dog, yielding to adverse star, sets," for example, refers to springtime, when the sun was in Taurus between April and May. At this time of the year Sirius ("the dog") comes closer to the sun and becomes invisible. From these cues the timings for the planting, care, and harvesting of each plant can be gauged.

Heliacal risings were also used for this kind of timekeeping. They appear when the celestial object's rising is simultaneous with the sun's rising. The star or planet is visible for a very short period and then hidden by the sun's brightness. Heliacal settings of stars or planets, on the other hand, happen if the object's setting is simultaneously to the sun's rising. In this case the object is visible for a short time before it sets (Lieven 2007, 144).

The *Fasti* by Ovid deals with the Roman calendar. Each chapter belongs to one of the months of the year. The quotation "now aged Aquarius sits below with urn inclined," Sacrobosco cites is from Book II and refers to February 15. On this day, as Ovid describes, Aquarius is hidden by the nearness of the sun while it moved to Pisces. In ancient Rome February 15 was celebrated as *Lupercalia*, a pastoral festivity (*Encyclopedia Britannica* 2021).

Sacrobosco's use of the poets' works gives a clear indication of the connection of celestial phenomena to certain events of the year in the ancient world to keep track of the seasons, festivities, and months. At the beginning of Chapter III Sacrobosco even refers to a difference between interpretations of risings and settings of the signs by poets and by astronomers. The heliacal volvelle clearly refers to the poetic interpretation, as it is introduced with the phrase *Instrumentum, quo facilime omnes diversitates ortus Poetici, oculis subijciuntur* ("The instrument, by which all different poetical risings are brought to mind").

As the historian James Evans describes in his work on ancient astronomy, this kind of timekeeping and understanding of the annual cycle of star phases was a common practice in early Babylonian and early Greek astronomy. For example, the heliacal rise of the Pleiades for the first time of the year in late May "signaled the Wheat harvest and the beginning of summer weather" (Evans 1998, 190). He noted that these practices were systematized into a star calender called a *parapegma* (Evans 1998, 190). This is a time-keeping instrument in the form of a peg calendar that allows a person to keep track of the dates throughout the year by moving the pegs to different locations through the calendar (Fig. 6.15) (Rüpke 2006).

With the help of the heliacal volvelle students probably learned to work with the previously described connections and got to know the context of risings, settings, and different phases of the year.

As the letters on the second disc (b) are described neither in the text nor in the volvelles' instructions, their meaning is unclear. As Owen Gingerich writes, it is possible that they were used by university professors to create a particular setting of stars or constellations which they would work through together with the students (Gingerich 1981, 6).

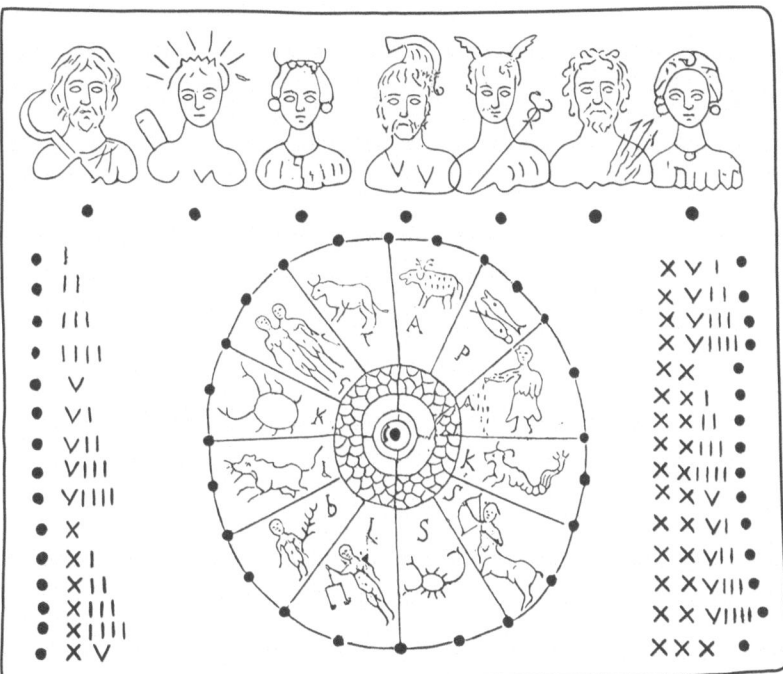

Fig. 6.15 This reconstruction shows an example of a *parapegma* from the fourth century CE that was found in a house close to the Baths of Trajans in Rome. It shows the sun's motion through the zodiac signs by means of moveable pegs. On the upper part of the calendar the seven Roman gods that are connected to the days of the week (from left: Saturn, Sun, Moon, Mars, Mercury, Jupiter, Venus) can be marked by a peg. The Roman numerals from I to XXX stand for the thirty days of a month, while the three holes per zodiac sign mark where the sun is located within the zodiac sign. From Wikimedia Commons, https://commons.wikimedia.org/wiki/File:Roman_calendar_-_parapegma_%28III_-_IV_c._C.E.%29.svg

6.4.2 Evaluation

The described heliacal volvelle is found in almost the same shape and size across the six investigated copies, as well as in the editions that are connected to the database of the Sphere project. The *parapegma* shown in Fig. 6.15 has a similar structure, as does the zodiac volvelle. Nevertheless, this exact volvelle seems to be unique in the corpus of early modern astronomical books, as it has not been found in another book thus far. Even in the *Sphaera* it was excluded from the Wittenberg group in some editions.

6.4.3 Cultural Analysis

Besides the affiliation of this artifact with the corpus of *Sphaera* texts that shaped the scientific and cultural identity of Europe through its long commentary tradition, it also has a didactic function in providing a hands-on solution to the context of risings, settings, and annual phenomena. For students studying astronomy in an early modern university the heliacal volvelle was probably helpful in visualizing and memorizing the context of the sun, stars, and zodiac signs in order to acquire knowledge on annual agricultural practices or festivities. By turning the wheel inside the classroom and hearing about the agricultural events that were connected to a certain rising, students got to know a technique of timekeeping and interpreting annual cycles. They learned how the agricultural year matched with the astronomical year.

6.4.4 Interpretation

This volvelle and the underlying theory and knowledge concerning astronomical phenomena are still relevant to our culture, but in a broad sense. The modern calendar still follows the annual motion of the sun and belongs to the category of solar calendars. Thanks to modern methods and tools, and the widespread and accessible nature of time and calendar calculations, as well as weather forecasts, the connections between celestial events and agriculture no longer play a role for the respective people, such as farmers or astronomers. Farmers are guided by the recurring months and seasons, as well as the weather, provided through digital platforms when cultivating their fields.

The same is true of mobile annual religious festivities, such as Easter and Ramadan. These can be calculated with ease with the use of the solar calendar, records and computer programs. The connection between astronomy, agriculture, and festivities still exists, but is no longer visible or relevant to students of astronomy.

Consequently, the volvelle itself no longer plays a role for our present culture in opposite to the underlying knowledge which is still relevant as stated above.

Bibliography

Digital Repositories

Sphaera Corpus*Tracer* Max Planck Institute for the History of Science. https://db.sphaera.mpiwg-berlin.mpg.de/resource/Start.
USTC Universal Short Title Catalogue, https://www.ustc.ac.uk.

Archives

MPIWG Library of the Max Planck Institute for the History of Science, Berlin, Germany.
ONB Österreichische Nationalbibliothek (Austrian National Library), Vienna, Austria.

Primary Sources

Apianus, Petrus, and Johannes de Sacrobosco. 1526. *Sphaera Iani de Sacrobusto astronomiae ac cosmographiae candidatis scitu apprime necessaria per Petrum Apianum accuratissima diligentia denuo recognita ac emendata.* Ingolstadt: Petrus Apianus. http://hdl.handle.net/21.11103/sphaera.100070.

Blebel, Thomas. 1576–1577. *De sphaera et primis astronomiae rudimentis libellus ad usum scholarum triuialium, ut uocant, maximè accommodatus, & in gratiam studiosae iuuentutis Curianae, ex artificum libris accurata methodo & breuitate conscriptus a M. Thoma Blebelio rudissimi Budiss: Ludiliterarij Curiensis collega.* Wittenberg: Matthaeus Welack. http://hdl.handle.net/21.11103/sphaera.101306.

Giuntini, Francesco, and Johannes de Sacrobosco. 1582. *La sfera del mondo, Di M. Francesco Giuntini, Dottore Theologo: col testo di M. Giovanni Sacrobosco. Opera utile & necessaria à poeti, historiografi, naviganti, agricoltori, & ad ogni sorte di persone.* Lyon: Symphorien Béraud. http://hdl.handle.net/21.11103/sphaera.101118.

Melanchthon, Philipp, and Johannes de Sacrobosco. 1538. *Ioannis de Sacro Busto libellus, De sphæra: Eiusdem autoris libellus, cuius titulus est Computus, eruditissimam anni & mensium descriptionem continens. Cum praefatione Philippi Melanth. & novis quibusdam typis, qui ortus indicant.* Wittenberg: Joseph Klug.

Melanchthon, Philipp, and Johannes de Sacrobosco. 1545. *Ioannis de Sacro Busto libellus de sphaera. Accessit eiusdem autoris computus ecclesiasticus, Et alia quaedam in studiosorum gratiam edita. Cum praefatione Philippi Melanthonis, Libellus Computus.* Vitebergae: Veit Creutzer. http://hdl.handle.net/21.11103/sphaera.100818.

Peucer, Kaspar. 1558. *Elementa doctrinae de circulis coelestibus et primu motu.* Vitebergae: Crato. http://hdl.handle.net/21.11103/sphaera.100191.

Peucer, Kaspar. 1576. *Elementa doctrinae de circulis coelestibus, et primo motu, recognita et correcta. Autore Casparo Peucero.* Wittenberg: Johann Krafft the Elder. http://hdl.handle.net/21.11103/sphaera.100473.

Sacrobosco, Johannes de. 1559. *Sphaera Ioannis de Sacrobosco.* Antwerp: Jean Richard. http://hdl.handle.net/21.11103/sphaera.101107.

Secondary Sources

Defeyter, Margaret Anne, Riccardo Russo, and Pamela Louise McPartlin. 2009. The picture superiority effect in recognition memory: A developmental study using the response signal procedure. *Cognitive Development* 24 (3): 265–273. https://doi.org/10.1016/j.cogdev.2009.05.002.
Drennan, Anthony S. 2012. The bibliographical description of astronomical volvelles and other moveable diagrams. *The Library* 13 (3): 316–339. https://doi.org/10.1093/library/13.3.316.
Encyclopedia Britannica. 2020. Ursa major. https://www.britannica.com/place/Ursa-Major.
Encyclopedia Britannica. 2021. Lupercalia. https://www.britannica.com/topic/Lupercalia.
Evans, James. 1998. *The history and practice of ancient astronomy*. New York/Oxford: Oxford University Press.
Georges, Karl Ernst. 1918a. *occasus*. Vol. 2. Band, *Ausführliches lateinisch-deutsches Handwörterbuch*. Hannover. http://www.zeno.org/nid/20002526700.
Georges, Karl Ernst. 1918b. *ortus*. Vol. 2. Band, *Ausführliches lateinisch-deutsches Handwörterbuch*. Hannover. http://www.zeno.org/nid/20002526700.
Gingerich, Owen. 1981. Astronomical scrapbook. Early textbooks with moving parts. *Sky & Telescope* 61: 4–6.
Gingerich, Owen. 1993. Astronomical paper instruments with moving parts. In *Making instruments count. Essays on historical scientific instruments presented to Gerard L'Estrange Turner*, ed. Robert G.W. Anderson, Jim Bennett, and William F. Ryan, 63–74. Aldershot: Variorum.
Hamel, Jürgen. 2014. *Studien zur "Sphaera" des Johannes de Sacrobosco, Acta historica astronomiae*. Leipzig: AVA, Akademische Verlagsanstalt.
Helfand, Jessica. 2002. *Reinventing the wheel*. New York: Princeton Architectural Press.
Kanas, Nick. 2005. Volvelles! Early paper astronomical computers. *Mercury* 34 (2): 33–39.
Karr, Suzanne. 2004. Constructions both sacred and profane: Serpents, angels, and pointing fingers in Renaissance books with moving parts. *The Yale University Library Gazette* 78 (3/4): 101–127.
Kräutli, Florian, Daan Lockhorst, and Matteo Valleriani. 2020. Calculating sameness: identifying early-modern image reuse outside the black box. *Digital Scholarship in the Humanities* 36 (Supplement_2): ii165–ii174. https://doi.org/10.1093/llc/fqaa054.
Lieven, Alexandra von. 2007. *Grundriss des Laufes der Sterne. Das sogenannte Nutbuch*. Copenhagen: Museum Tusculanum Press.
Limbach, Saskia. 2022. Scholars, printers, and the sphere: New evidence for the challenging production of academic books in Wittenberg, 1531–1550. In *Publishing Sacrobosco's De sphaera in early modern Europe: Modes of material and scientific exchange*, ed. Matteo Valleriani and Andrea Ottone, 147–185. Cham: Springer.
Meyers Großes Konversationslexikon. 1908. Mittag. In *13. Band*. Leipzig/Wien: Verlag des Bibliographischen Instituts.
Peters, Christian Heinrich Friedrich, and Edward Ball Knobel. 1915. *Ptolemy's catalogue of stars. A revision of the Almagest*. Washington, DC: The Carnegie Institution of Washington.
Rideau-Kikuchi, Catherine. 2022. Erhard Ratdolt's Edition of Sacrobosco's Tractatus de sphaera: A new editorial model in Venice? In *Publishing Sacrobosco's De sphaera in early modern Europe: Modes of material and scientific exchange*, ed. Matteo Valleriani and Andrea Ottone, 61–98. Cham: Springer.
Rosen, Jill. 2017. Memorable moves: How an object moves affects how well we recall it. John Hopkins University Hub. https://hub.jhu.edu/2017/03/06/object-movement-affects-memory/.
Rüpke, Jörg. 2006. Parapegma. *Der neue Pauly*. http://referenceworks.brillonline.com/entries/der-neue-pauly/parapegma-e907950.
Schurgin, Mark W., and Jonathan I. Flombaum. 2017. Exploiting core knowledge for visual object recognition. *Journal of Experimental Psychology: General* 146 (3): 362–375. https://doi.org/10.1037/xge0000270.
Thorndike, Lynn. 1949. *The sphere of Sacrobosco and its commentators*. Chicago, IL: The University of Chicago Press.

Valleriani, Matteo, Beate Federau, and Olya Nicolaeva. 2022. The hidden praeceptor: How Georg Rheticus taught geocentric cosmology to Europe. *Perspectives on Science* 30 (3): 407–436. https://doi.org/10.1162/posc_a_00421.

Open Access This chapter is licensed under the terms of the Creative Commons Attribution 4.0 International License (http://creativecommons.org/licenses/by/4.0/), which permits use, sharing, adaptation, distribution and reproduction in any medium or format, as long as you give appropriate credit to the original author(s) and the source, provide a link to the Creative Commons license and indicate if changes were made.

The images or other third party material in this chapter are included in the chapter's Creative Commons license, unless indicated otherwise in a credit line to the material. If material is not included in the chapter's Creative Commons license and your intended use is not permitted by statutory regulation or exceeds the permitted use, you will need to obtain permission directly from the copyright holder.

Chapter 7
Conclusion

Abstract This chapter describes the outcome of the study of moveable paper wheels, so called volvelles, within the *Sphaera* corpus examined by the research project The Sphere at the Max Planck Institute for the History of Science. The volvelles within the corpus were used for pedagogical purposes and as material models to display movements of spherical objects and underlying astronomical concepts and knowledge. For this, these moveable paper wheels made use of the picture superiority effect, which refers to the brain's tendency to memorize (moveable, in this case) images more easily than words. Additionally, it was shown how the largest group of volvelles, initially printed in Wittenberg, was reprinted and spread all over Europe, which supports the thesis of an epistemic community existing in Wittenberg. Finally, it was proven that 34% of the known *Sphaera* editions hold volvelles, which makes the volvelles an important part of the book's tradition.

Keywords *Sphaera* · Johannes de Sacrobosco · Volvelles · Paper instruments · Material culture · Picture superiority effect · Early modern period · History of science · History of knowledge

In the present work the context of the volvelles of Johannes de Sacrobosco's *Sphaera* has been elaborated using a corpus of representative copies of the Sphere and with the help of historical, museal, and digital methods.

In Chapter 1 it was stated that different methods, such as the investigation of historical artifacts, can enrich the historical research experience. It was asked if the combination of text, image, and instrument can offer other insights into premodern scientific knowledge and thought than can just the artifact or just the text.

Chapter 2 gave an introduction to astronomical instruments to show how in early modern times crucial practical astronomical knowledge such as navigation, time-measuring, and calculation was incorporated in instruments such as the *gnomon* or the astrolabe. In particular, instruments such as the astrolabe used the so called "stereographic projection," a mapping method that transfers a three-dimensional spherical surface to a two-dimensional plane. This knowledge is important in understanding how astronomical instruments work, and was a technique also applied to paper instruments, to which genre volvelles belong. Paper instruments were an important tool for hands-on practice and also served as entertainment in the form of pop-up geometrical figures, flaps, or separate kits for building scientific instruments or volvelles. This medium taught readers to read images as interactively as they read texts and opened up a world of mathematical, astronomical, and medical knowledge and practice, especially for those who lacked access to actual instruments made from expensive materials such as brass. Volvelles had a kind of "golden age" shortly after the introduction of printing in Europe in the fifteenth century. Authors such as Ramon Lull and Peter Apian influenced this genre widely through the incorporation of moveable instruments in their works. According to Suzanne Karr Schmidt, about 190,000 interactive printed copies must have been circulating in the sixteenth century alone.

The third chapter examined the context of the *Sphaera* by Johannes de Sacrobosco. Sacrobosco's lifetime and the time of the creation of the *Sphaera* was an important period for astronomy, which became part of the seven liberal arts at the early modern European university. The *Sphaera*'s elementary character as well as its content made it among the most used textbooks in astronomy at that time. The book's content was presented in four chapters, describing the properties of spheres; the rising and setting of signs; climes; motions of planets and eclipses. Reading a book in medieval and early modern times was a practical and interactive matter, and from this close interaction a commentary tradition emerged in which new or complementary knowledge was added to book such as the *Sphaera*. These additions were made in the form of new text between, around or on the margins of the original text. As a result of this practice the *Sphaera* underwent a transformational process over the centuries: knowledge concerning mathematical astronomy, calendrical calculations, the use of and instructions for constructing astronomical instruments, nautical astronomy and geography, cartography, meteorology, arithmetic, geometry, judiciary astrology, literature, practical objects, and mechanics was added. To include other topics and form a volume that contained extended knowledge, the *Sphaera* was often bound together with works such as the *Theorica planetarum* by Gerard of Cremona or the *Theorica novae planetarum* by Georg Peuerbach. The way in which the knowledge connected to the *Sphaera* evolved and transformed shaped the scientific identity of Europe, particularly as, as Owen Gingerich has calculated, more than 200,000 copies of the *Sphaera* must have been printed during the early modern period.

The research project The Sphere at the Max Planck Institute for the History of Science in Berlin collected 359 editions of the work, a number that suggests that Gingerich's calculations were an under-estimate: At least 350,000 copies must have been circulating between 1472 and 1650—the targeted timespan of the research project. In the frame of the project a database was created that stores one copy per *Sphaera* edition together with a record of all possible information on it—the author, printer, publishing date and place, content, images etc. In particular, the images are still being investigated to explore the visual culture around the *Sphaera*. More than 20,000 images were collected and captured by hand and with the help of a machine learning algorithm. With the help of this image capturing process, it was possible to extract all volvelle parts from the database and cluster them using the Sphaera

7 Conclusion

Infrastructure Tool to form a research corpus. The copies that were collected for the project's corpus contain 593 pieces of volvelles, including completely assembled volvelles and single parts. I used this set to examine the use and function of the volvelles inside the *Sphaera* corpus.

* * *

Volvelles appeared in *Sphaera* editions from the sixteenth century on: Joseph Klug included the first volvelle in an edition printed in Wittenberg in 1538. From then, volvelles can be found in *Sphaera* works until 1647, when the last edition with volvelles was printed in Leiden by Bonaventura and Abraham Elsevier. During these 109 years about 123 *Sphaera* editions that included volvelles were printed, which makes up 34% of the 359 *Sphaera* editions printed from 1450 to 1672. This suggests that around 123,000 copies of the *Sphaera* with volvelles existed.

Suzanne Karr Schmidt has argued that about 190,000 copies of works with moveable parts must have existed in sixteenth-century Europe. However, as I was able to find 98 *Sphaera* editions with moveable parts from the sixteenth century alone, implying around 98,000 copies of *Sphaera* works with moveable parts alone, it seems plausible to venture that there must have been many more books with moveable parts in the sixteenth century than Karr Schmidt estimated.

I identified twelve different kinds of volvelle inside the corpus. These are not as complex as other volvelles from the early modern period, such as those from Apianus' *Astronomicum Caesareum*. As neither the volvelles nor their making were the main focus of the *Sphaera*, it should be asked why they were included in the *Sphaera*. In part, they are undoubtedly a result of the aforementioned "golden age" of paper instruments within the printing industry that began shortly after the introduction of printing to Europe in 1450. Secondly, moveable parts, including volvelles, had different functions: They were included for entertainment, for didactic or for mnemonic purposes, through visualizing or repeating basic concepts.

The twelve volvelles from the corpus can be classified in three groups: the Wittenberg group (4 volvelle types), the Leiden group (1 volvelle type), and the Seville group (6 volvelle types). I briefly described the volvelles from the Leiden and the Seville groups before analyzing the volvelles from the Wittenberg group in detail. The Leiden group contains just a single type of volvelle, which was printed in four editions in Leiden by Bonaventura and Abraham Elsevier in 1626, 1639, and 1647. The Seville group consists of six types of volvelle that were probably useful for navigational purposes as they appeared in Martin Cortés's work *The Arte of Navigation*, which was an adaption of the original *Sphaera* text. This group of volvelles was printed in Seville in 1551 and remained part of the printing tradition for 79 years, being printed for the last time in London in 1630. The Wittenberg group consists of four volvelles and accounts for the largest share of the corpus: the volvelles from this group appear in 109 *Sphaera* editions between 1538 (printed in Wittenberg) and 1639 (one from Salamanca, two from Wittenberg), which make up 89% of all editions containing volvelles. Though the number and conditions of volvelle pieces from the project's database are not representative of the whole corpus of *Sphaera* copies from the early modern period, I investigated this data to give an indication of the culture around volvelles, especially those of the Wittenberg group.

As historians such as Limbach and Rideau-Kikuchi have shown, it was quite common for early modern printers to share their woodblocks with each other or to copy them (Limbach 2022; Rideau-Kikuchi 2022). Owing to this practice, so called "epistemic communities" of the early modern period could easily be traced and it was shown by Zamani et al. (2020) that Wittenberg was a center of such an epistemic community, where innovations were fostered and spread all over Europe. The findings from the underlying volvelle corpus supports this thesis as well: Indeed, books of this group were most often printed in Wittenberg (42 editions) but from there spread across Europe via Paris (27 editions) and Antwerp (13 editions) in 1543. It is also important to consider legal factors, such as printing privileges, when discussing knowledge diffusion and epistemic communities.

I assigned terms to the four volvelle types in the Wittenberg group based on their apparent functionality: the eclipse volvelle, the horizon volvelle, the zodiac volvelle, and the heliacal volvelle. In most cases the whole group of four can be found inside an edition or copy, while in some cases the zodiac volvelle appears alone in an edition, and the heliacal volvelle was sometimes excluded from the group.

The volvelles appear in the corpus in different conditions: They can be assembled, correctly or incorrectly; they can be disassembled; or they can be in their original condition, with no attempt made at all to assemble them. In the last case, the base disc still exists inside the book on the page where the volvelle is meant to be assembled.

Base discs form the largest group of volvelle pieces in the Wittenberg group and in the *Sphaera* corpus, at around 57%. Together with other single parts, they make up 77% of the volvelle pieces in the corpus, in the form of "not assembled volvelles." Base discs in particular can tell us about the number of editions containing volvelles, because they were integral to the physical book and did not become mislaid, as parts did. For the same reason they also show in which part of the text the volvelle should have been installed. In instances where volvelles were not assembled, the appearance of the base disc is evident, as it is not hidden by other volvelle parts.

Parts of volvelles were often lost if they weren't assembled. Those parts still extant show how they were printed and sold to be assembled by a reader: The sheets with parts were printed on a folio sheet and then folded three times to fit into a book of the octavo format. They were sold and bound together with the rest of the book, but as a separate kit in the back of the book. The reader had to cut them out and install them inside the book on the designated page. As these sheets with parts still exist in some cases, it can be seen that they were sometimes printed together with instructions. The instructions tell the reader how to assemble all of the four volvelles inside the book, using threads, although readers sometimes had more elaborate ideas and crafted paper pins to make the volvelle easier to handle and longer lasting. I also found editions where the parts were printed and bound in a way that hindered the cutting and assembling process. Either they were printed on a back of a page containing part of the main text or the sheet was folded and bound like the other pages, and thus the parts were halved. This suggests that printers and bookbinders lacked an understanding and knowledge of the volvelles' assembly or purpose, or

7 Conclusion

perhaps felt it unlikely that the reader would actually assemble them and that special treatment was therefore unnecessary.

Though very few sheets exist that contain instructions on how to assemble the volvelles, the volvelles of the Wittenberg group were often assembled correctly, when they were assembled at all. Correctly assembled volvelles make up 19% of the Wittenberg group. The zodiac and horizon volvelles were most often assembled correctly. They were also the least complex volvelles, containing of just two (zodiac volvelle) or three (horizon volvelle) parts. The heliacal volvelle was more often cut or assembled incorrectly. It seemed to be unclear to readers how to cut the extra squares on the second disc (b). However, as volvelles were most often assembled correctly, when assembled, it is possible that students discussed how to assemble the volvelles in a correct manner.

To understand the material culture of the Wittenberg group I utilized Edward McClung Fleming's Winterthur model and Davis Baird's concept of Thing Knowledge. The Winterthur model, which follows a certain workflow involving the stages of Identification, Evaluation, Cultural analysis, and Interpretation, was applied to six cases of each of the four volvelles and delivered information on all of the four types of volvelles. Thing knowledge follows the concept that both artifacts and theory can compose our knowledge of the world, while instruments are a kind of scientific knowledge. Theory and object have an epistemological equivalence, although there is no unified epistemological treatment for both. The theory underlies the idea that every instrument works epistemologically differently concerning the knowledge it constitutes. As far as the *Sphaera* volvelles are concerned, Baird's idea of "models" is the most relevant.

The eclipse volvelle appears in the first chapter of the *Sphaera*. It was probably intended to demonstrate how an eclipse can be seen with respect to the horizon and how it appears in different regions of the world at different times. According to Baird, the eclipse volvelles would be a "model," containing model knowledge that is explanatory, predictive, and simple. The knowledge that is contained in this volvelle was previously explained through the text and thus the volvelle was not crucial to understand the phenomenon. However, the picture superiority effect explains its usefulness to students, who were probably able to understand the phenomenon more easily by manipulating the volvelle. Other volvelles of this kind often look similar, if not identical, but they are limited to the *Sphaera* corpus and do not appear in other early modern books.

The horizon volvelle also appears in the first chapter of the *Sphaera*. This volvelle, as well as the heliacal volvelle, were in the connected texts of the *Sphaera* works referred to as "instruments." This one shows the earth's curvature in a latitudinal way by visualizing the paths of the sun in respect to the horizon and certain star constellations. Together with the eclipse volvelle, a ship diagram from the first chapter and the surrounding text, it demonstrates the spherical shape of the earth. The horizon volvelle also belongs to the category of a "model." It enriched the chapter by making an understanding of the spherical shape of the earth easier to grasp and the celestial movements easier to remember. This volvelle is similar to the volvelle from the

Leiden group and to the first volvelle from the Seville group. Neither this nor the previous volvelle have relevance in today's culture.

The zodiac volvelle appears in the second chapter. It is a model of the zodiac circle, and is described in the surrounding text in detail. According to Baird's theory, the zodiac volvelles would also be "models," containing model knowledge that is explanatory, predictive, and simple. The concept explored through the volvelle is explained in the text and thus the volvelle was not crucial to understanding the phenomenon, but it was probably very helpful for mnemonic reasons. This kind of volvelle and also the base disc as a diagram was very common in early periods, and the basic concept is still used today as a visualization of the zodiac circle for astrological purposes.

The heliacal volvelle appears in the third chapter. It is, like the horizon volvelle, in the *Sphaera* texts described as an instrument. This volvelle helps readers to understand the connection between celestial events and annual seasons or festivities. Students probably learnt to calculate these connections and work with them with the help of the volvelle. The model knowledge from this instrument is still relevant for agriculture or calendars, but the volvelle itself is not.

All the volvelles except the eclipse volvelle work with the zodiac signs and are assembled in the same or a similar manner by threading and gluing the parts to the designated page. They all have an individual function and meaning, but were generally meant as material models to support the students in memorizing and understanding the matter in the surrounding text. Jürgen Renn describes a "'model' as a corresponding external knowledge representation structure" (Renn 2015). Material models are a representation of a corresponding mental model; so, in the case of these volvelles, the zodiac volvelle represents the mental model of the sun's movement through the zodiac signs, for example. This material model (the volvelle) cannot substitute the mental model, but it makes it possible for an individual (the student) to participate and learn the shared knowledge that underlies the models. Additionally, the material model could include features that "had not been included in its initial conception as the materialization of a mental model" (Renn 2020, 78). The benefit of learning with the help of volvelles is also supported by modern concepts such as the "picture superiority effect," described above (Rosen 2017; Schurgin and Flombaum 2017). Therefore, the students were supported by the volvelles in memorizing, for instance, the heavenly movements and the order of the zodiac signs, gaining a more plastic understanding of the text contents in the process.

Why volvelles disappeared from the *Sphaera* texts is still unknown. Eventually, the content of the *Sphaera* was more and more supported by additional commentaries or images throughout the centuries, which might account for the volvelles becoming increasingly irrelevant. Another possibility is that images or instruments evolved in a way that made volvelles unnecessary to comprehend the content, or stand-alone paper instruments that were not incorporated in books were becoming cheaper and more widespread, replacing volvelles.

To provide further support for the work done here, an analysis of the other volvelle groups, but especially of the Seville group, would be a research desideratum. Additionally, it would be interesting to explore more *Sphaera* copies and gain more data and information about the distribution of assembled volvelles, to ascertain whether failing to assemble volvelles was common practice.

This work is the first to examine the movable parts of a centuries-old book and knowledge tradition of the early modern period—the volvelles of Johannes de Sacrobosco's *Sphaera*. For the first time, a detailed examination of the functions of the individual volvelles was carried out and their use was presented with the help of a model in order to show their connection to the knowledge contained in the *Sphaera* texts. Most likely, volvelles were included in the *Sphaera* for didactic reasons, to help students understand more complex sequences and movements of celestial bodies using the so called *Picture Superiority Effect*.

In addition, the present study proves that there must have been more Sphaera works with volvelles than previously assumed, and also many more books within the Sphaera corpus with movable parts than previously thought. In addition, it was possible for the first time to identify all the volvelles in the Sphaera corpus and to carry out a quantitative data analysis of their occurrence. The *Sphaera* corpus of printed editions contains significantly more editions with volvelles than expected: 34% of the total. Over the nearly 200 years of the printed tradition of the Sphaera, this figure is a significant proportion of the total amount of Sphaera works. This figure suggests that the volvelles are a more important aspect of Sphaera culture than previously thought. In addition, a brief excursion into the legal systems surrounding the publication of books and thus the distribution of knowledge in the form of volvelles in the early modern period was presented and discussed under the topic of *privilegia impressoria*. Considering that printing privileges were an important tool for printers, publishers and authorities, they should be included in the research on the factors of knowledge distribution in early modern Europe. This leads to further questions that should be investigated in the future, e.g. what role local jurisdiction played in the production and dissemination of written knowledge. The study of the underlying socio-economic and legal mechanisms of knowledge distribution can ultimately contribute not only to the understanding of the European privilege system in the French and German jurisdictions, but also to the understanding of certain epistemological trends, including volvelles.

Bibliography

Secondary Sources

Limbach, Saskia. 2022. Scholars, printers, and the sphere: New evidence for the challenging production of academic books in Wittenberg, 1531–1550. In *Publishing Sacrobosco's De sphaera in early modern Europe: Modes of material and scientific exchange*, ed. Matteo Valleriani and Andrea Ottone, 147–185. Cham: Springer.

Maryam, Zamani Alejandro, Tejedor Malte, Vogl Florian, Kräutli Matteo, Valleriani Holger, Kantz. 2020. Evolution and transformation of early modern cosmological knowledge: A network study Abstract Scientific Reports 10(1). https://doi.org/10.1038/s41598-020-76916-3.

Renn, Jürgen. 2015. From the history of science to the history of knowledge—And back. *Centaurus* 57 (1): 37–53. https://doi.org/10.1111/1600-0498.12075.

Renn, Jürgen. 2020. Chapter 4: Structural changes in systems of knowledge. In *The evolution of knowledge*, ed. Jürgen Renn, 65–86. Princeton, NJ: Princeton University Press.

Rideau-Kikuchi, Catherine. 2022. Erhard Ratdolt's edition of Sacrobosco's Tractatus de sphaera: A new editorial model in Venice? In *Publishing Sacrobosco's De sphaera in early modern Europe: Modes of material and scientific exchange*, ed. Matteo Valleriani and Andrea Ottone, 61–98. Cham: Springer.

Rosen, Jill. 2017. Memorable moves: How an object moves affects how well we recall it. John Hopkins University Hub. https://hub.jhu.edu/2017/03/06/object-movement-affects-memory/.

Schurgin, Mark W., and Jonathan I. Flombaum. 2017. Exploiting core knowledge for visual object recognition. *Journal of Experimental Psychology: General* 146 (3): 362–375. https://doi.org/10.1037/xge0000270.

Open Access This chapter is licensed under the terms of the Creative Commons Attribution 4.0 International License (http://creativecommons.org/licenses/by/4.0/), which permits use, sharing, adaptation, distribution and reproduction in any medium or format, as long as you give appropriate credit to the original author(s) and the source, provide a link to the Creative Commons license and indicate if changes were made.

The images or other third party material in this chapter are included in the chapter's Creative Commons license, unless indicated otherwise in a credit line to the material. If material is not included in the chapter's Creative Commons license and your intended use is not permitted by statutory regulation or exceeds the permitted use, you will need to obtain permission directly from the copyright holder.

The manufacturer's authorised representative in the EU is Springer Nature Customer Service Centre GmbH, Europaplatz 3, 69115 Heidelberg, Germany. If you have any concerns regarding our products, please contact ProductSafety@springernature.com

Printed and bound by CPI Group (UK) Ltd, Croydon, CR0 4YY

26/03/2026

02078953-0013